Optimisation of nutrient cycling and soil quality for sustainable grasslands

Optimisation of nutrient cycling and soil quality for sustainable grasslands

Proceedings of a satellite workshop of the XXth International Grassland Congress, July 2005, Oxford, England

edited by:
S.C. Jarvis
P.J. Murray
J.A. Roker

Subject headings:
Soil biodiversity
Soil physical conditions
Soil chemical interactions

ISBN 9076998728

First published, 2005

Wageningen Academic Publishers
The Netherlands, 2005

The Institute of Grassland and Environmental Research
North Wyke Research Station
Okehampton, Devon, EX20 2SB, UK

Acknowledgements

We thank colleagues at North Wyke Research Station for their various contributions to the organisation and running of this Workshop. IGER is sponsored by the Biotechnology and Biological Research Council (BBSRC) who also provided additional support. We would also like to acknowledge other support from the Department of Food, Environment and Rural Affairs (Defra), London and Kemira GrowHow, UK.

Workshop steering committee

S.C. Jarvis	IGER
P.J. Murray	IGER
J.A. Roker	IGER
R. Bardgett	Lancaster University
J. Crichton	British Grassland Society
D. Fay	Teagasc
I. Richards	Ecopt

Foreword

This book is the published outcome of the IGC satellite workshop on 'Optimisation of nutrient cycling and soil quality for sustainable grasslands' held at Saint Catherine's College, Oxford, July 2005.

The objective was to attempt to bring together two aspects of grassland soil management which, by and large, have hitherto been considered separately. Issues related to nutrient cycling and soil quality have dominated research directed towards aiding broad and local scale policy issues for improving land use, protecting the environment and maintaining/preserving natural habitats and biodiversity, but have tended to have considered separately. In this book we attempt to bring what are, in reality, inseparable aspects of grassland soil characteristics together and consider their physical, chemical and biological components, their interrelations and the way that they influence nutrient transformations and flows and soil quality. For each component an invited lead speaker opened the discussions which included both oral and poster presentations. Whilst we have placed the various contributions in discreet sections, it was very clear that segregating the information in this way is very artificial and there was enormous overlap and opportunity for future integration between the different sectors of the area.

The opportunity to bring together international expertise and experience will do much, we hope, to progress understanding and point ways forward to maintain and sustain what is a base resource, our soils. This is essential whether it be for production targets, environmental benefit or for maintenance of natural ecosystems, in good order for future generations. The book should be of interest to all those interested in soils and their function *per se*, and to all grassland managers, whether their aims are directed at producing food, forage or fibre of sustainable quantity and quality or at maintaining, restoring or encouraging above and below ground biodiversity. The international perspective on this is very important so that experiences in wide ranging circumstances can be cross-referenced and used to the advantage of all.

The Editors, July 2005

Table of contents

Keynote presentations

Soil biology and the emergence of adventive grassland ecosystems

T.R. Seastedt

Department of Ecology and Evolutionary Biology, University of Colorado, Boulder, Colorado 80309, USA, Email: timothy.seastedt@colorado.edu

Key points

1. Understanding the role of biodiversity in ecosystem function has dominated the research efforts of grassland ecologists in recent years. Research indicates that the trophic complexity of the soil biota precludes simple predictive responses to manipulations of intact soil systems.
2. The composition and activity of the soil biota are sensitive to environmental drivers and changes in plant species composition. The emergence of adventive ecosystems, generated by introduced species and new environmental conditions of grasslands, demands a research emphasis to assess the ecosystem services provided by the soil biota while concurrently addressing species conservation concerns.

Keywords: biodiversity, environmental change, invasive species, nitrogen deposition

Introduction

The last few years have seen an explosion of information on the responses and feedbacks of soil biota to species and environmental changes. Soil biologists, plant ecologists, and biogeochemists have been searching for patterns, generality and predictability in these relationships. A few brave souls have had the courage to assemble comprehensive assessments of soil biology in our rapidly changing world. Several years ago, I reviewed Wardle's (2002) book emphasising the links between above- and belowground components of ecosystems. That activity – as well as the present one – provided me with the opportunity to survey recent findings in soil biology. I have attempted to synthesise that information with respect to conservation and management issues. This analysis suggests that our scientific endeavours continue to advance the goal of providing sustainable, productive grasslands. At the same time, nevertheless, I sense growing urgency to identify the consequences of inevitable abiotic and biotic change on the structure and function of grassland soil ecosystems.

The current research foci of grassland ecologists

To assess recent research emphasis among grassland ecologists, I conducted a subject-directed literature search for 2001- 2004. The Science Citation Index (Web of Science) was used to assess grassland research topics published in two journals each in North America (*Ecology* and *Ecological Applications*), the United Kingdom (*Journal of Ecology* and *Journal of Applied Ecology*) and mainland Europe (*Oecologia* and *Oikos*). Topics included 1) climate change, 2) carbon dioxide, 3) nitrogen deposition, 4) invasions, 5) land use, 6) historical range of variability (disturbance, fire, and floods) and 7) biodiversity. The effort was iterated several times using various word combinations to provide some confidence that the analysis reflects activity on the research topics. Of the total 4333 papers accessed, 2039 were from the mainland European publishers, 1600 from North America, and 694 from the UK. From this analysis, grassland publications on the above topics totalled 77, 79, and 58, in these sources, respectively. In terms of percentage of total publications, the UK journals have the greatest emphasis on grassland topics. Grassland citations represented 4%, 5%, and 8% of total publications in mainland Europe, USA, and UK, respectively.

Clearly, the amount of science directed at grasslands is disproportionately low for the relative global coverage of grasslands. Grassland publications represented only about 5% of the subject area of the 4333 papers searched in this effort. For comparison, I repeated this exercise a few months later using the terms 'forest' and 'desert' in the same journals and found 1231 of 4705 citations (*ca* 26%), on these topics, mostly on forests. This discrepancy may be made up in speciality journals emphasising soils, soil ecology, and agroecosystems, but it does show a lack of emphasis on grassland studies *per se*. A search for topics on 'desertification', 'forestation or deforestation', 'grassification' and 'grasslandification' provides additional proof of this bias. Humans have been in the business of creating grasslands and grazing lawns for centuries (or 'grassified', see Baker, 1978), but the process of growing grass (as opposed to growing trees) is defined by what is lost, not what is gained. The term 'grasslandification' is not recognised except in those regions where invasive grasses are now replacing shrubland and desert flora.

Collectively, authors of grassland studies see biodiversity as the dominant research agenda, with 42% of grassland papers addressing this issue (Figure 1). The topic was particularly dominant in European publications, where the subject occupied almost 60% of grassland publications. This emphasis was reinforced when these same topics were searched using a broader array of ecology journals. Disturbance ecology and invasions were of secondary importance. The global change drivers of climate, carbon dioxide (CO_2), and nitrogen (N) deposition, somewhat surprisingly accounted for <10% of publications. The ecological community's interest in effects of external ecosystem drivers is overwhelmed by interest in internal dynamics. For example, when the phrase "nitrogen deposition" or "nitrogen input" is shortened to "nitrogen" in the search, citations increased from 13 to 599 citations from a search of 4705 papers. Causes and consequences of nitrogen availability in grassland soils remain a key interest to ecologists.

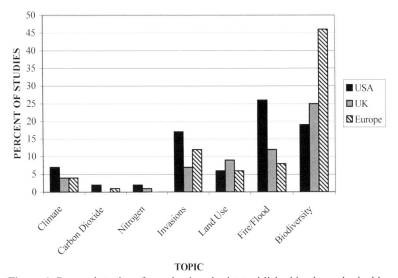

TOPIC

Figure 1 Research topics of grassland ecologists published in six ecological journals, 2001-2004

Differences might be expected between the research foci of scientists occupying a continent that has been under extensive human dominance for many centuries, and one that has been transformed relatively recently. North American publications emphasise change in historical ranges of variability, biodiversity and invasive species, with about equal emphasis on all three topics (Figure 1). The reduced emphasis on biodiversity *per se* makes sense in that, relative to the UK and Europe, many grasslands of North America have only recently emerged from relatively pristine conditions and are now outside their historical range of variability in terms of grazing, atmospheric chemistry, fire and flood return intervals. These transitions, with an emphasis on non-native species introductions, are perceived as both stand-alone subjects and variables affecting grassland biodiversity. Not surprisingly, this provincial bias is reflected in my analysis. My goal is to characterise the current status of most North American grasslands, and make what predictions are tenable about issues of sustainability and ecosystem services.

Advances in understanding the role of soil biodiversity

Since the publication of Wardle's (2002) analysis of controls and significance of soil biological systems, over 600 papers using the keywords of 'soil' and 'biodiversity' have appeared in the literature. Nevertheless, besides some stellar exceptions, our assessment of biodiversity remains, at most, fragmentary. Any attempt to sketch an authentic food web emphasises the complexity that biologists confront in the soil (Mikola *et al.*, 2002, Moore *et al.*, 2003). While manipulations of food chains above ground may produce trophic cascade effects, similar manipulations below ground produce less-predictable responses due to the true web-like complexity of the food web (e.g. Wardle & van der Putten, 2002; Moore *et al.*, 2004). Even if we understood this structure completely and further understood how different management scenarios altered this structure for a particular grassland, generalising how this structure would vary across resource gradients and how that variation translates into changes in the functional attributes of grassland ecosystems would be tenuous at best. While we may be able to identify our favourite species, and discuss how the abundance and activity of these organisms influence function, our ability to provide managers with an adequate assessment of the status and trends of soil biotic diversity and its consequences on function is not forthcoming.

Soil microbiologists and nematologists now conduct molecular phylogenic inventories to obtain censuses of bacterial, archaeal and selected eukaryal organisms. Regretfully, these tools have yet to proliferate to most soil biologists. The result is that we cannot tell the public and policy makers what is in the soil (Andre *et al.*, 2002), and, even if we could, explaining how this diversity translates to precise patterns and controls on ecosystem function would remain problematic. While there is now strong evidence that manipulations of soil biodiversity produce idiosyncratic relationships with ecosystem processes, studies of individual trophic groups or individual species almost always demonstrate positive or negative relationships (e.g. De Deyn *et al.*, 2003). What seems appropriate, therefore, is that we make every effort to conserve this diversity while we continue our attempts to identify what is there, what controls its presence and abundance, and what functions these species perform.

The key issue – the need to promote conservation while continuing the effort to provide baseline information – is itself an ongoing research issue. Studies and reviews provided by De Deyn *et al.* (2004), Wardle (2002, 2005), Wardle & van der Putten (2002) and Wardle *et al.* (2004) support the contention that resource abundance and heterogeneity are drivers of soil biodiversity. Soil biotic diversity can be facilitated by mechanisms that minimise plant competitive exclusion via herbivory (Bardgett & Wardle, 2003) and mechanisms that reduce dominance or competitive exclusion within the soil food web itself (Moore *et al*., 2003, 2004). Managing for high plant species diversity at local to landscape scales may therefore be the best *ad hoc* way to maximise soil biotic diversity.

Our research group (Porazinska *et al*., 2003; Reed, 2005; Wall *et al*., unpublished results) found strong evidence for this positive relationship between plant richness and soil species richness at the Konza Prairie Biological Station in Kansas, USA. Grass species attract unique assemblages of microbes attached to roots (Reed, 2004), as seen in similar studies of other herbaceous species (e.g., Callaway *et al*. 2004). Individual plant species therefore generate somewhat unique rhizosphere communities. These microbial and invertebrate communities impose feedbacks on the plants that can further enhance resource heterogeneity (e.g. Klironomos, 2003; Rudgers *et al*., 2004; Coleman *et al*., 2004). Nematode diversity in our study not only increased with plant species richness, but the patterning of plant species enhanced this diversity. A modest but statistically significant synergistic effect of species co-location was found on diversity (i.e. the nematode species richness beneath two co-located plant species was greater than the sum of the richness found when the same two plant species were sampled in isolation). We hypothesise that the combination of resource heterogeneity generated by different plant species, accompanied by temporal and spatial heterogeneity generated by root activities of different species (along with their different symbionts and pathogens), generate the highest diversity of organisms found in grasslands. This increase in species richness is generally consistent with numerical responses of herbivorous and microbivorous nematode functional groups (Todd, 1996, Blair *et al*., 2000). Finally, our group found that richness of nematodes appeared to be related to nutrient content or physiological status of the grass species. Richness was, on average, higher under cool season grasses (C_3 species) than warm season grasses (C_4 species). Collectively, the data are consistent with findings suggesting a large role of plant diversity in maintenance of soil biotic diversity. Given a general (but not universal) trend of C_3 species replacing C_4 species in North American grasslands, an increase in local richness of the invertebrate fauna might be expected.

Historical range of variability and the emergence of adventive ecosystems

The structure and function of grasslands have been changing since the first monocots began to dominate herbaceous communities many millions of years ago. Nonetheless, factors affecting resistance and resilience properties of these systems have been a focus of study for community ecologists for the last century. Ecologists have understood that discrete and sometimes rare events such as fire or floods affect ecosystems' structure and functioning. This concept matured to one where disturbances were observed as integral components of the system and referred to in the context of 'historical range of variability' or 'natural variability' (c.f. Landres *et al*., 1999). In North America, upland grasslands had historical fire and grazing return intervals that were strong determinants of species composition and structure, including the activities in soil biota. (e.g., Rice *et al*., 1998; Blair *et al*., 2000). This concept has valuable conservation management implications. For example, management to limit ungulate grazing in semiarid and arid natural areas in California and in the Great Basin of North America is based on the fact that most of these areas had few such grazers during the 12000-year interval prior to the early 19th century (c.f. Baker, 1978; Seastedt, 2002). Cattle grazing became the common scenario in this region, driving these ecosystems outside their historical range of variability. Such systems do not maintain their previous structural or functional traits. The now-altered system develops characteristics that may or may not be desirable in terms of ecosystem services, but is one clearly transformed from its previous state.

While fire and grazing intervals can be manipulated, atmospheric chemistry and climate change are not under land managers' control. Once historical ranges of important environmental drivers are exceeded, how will directional changes affect structure and function? We have been conducting this uncontrolled experiment on a global scale with atmospheric CO_2 and N concentrations. At the same time, human activities have been altering other environmental drivers at local and regional scales. New species, either in concert with environmental changes or independent of such changes, have also been added or subtracted from the system. What emerges is an altered or adventive ecosystem.

An adventive ecosystem is defined as an ecosystem containing native and non-native biological components, and exhibiting ecosystem properties resulting from a transition outside the historical range of variability (Figure 2). Causal mechanisms for the transition are attributed to both internal reorganisation and external drivers. While the devil is in the detail of this definition (and there exist a continuum of states), the term is meant to define systems that exist between those totally dominated by human forcings (i.e. synthetic ecosystems such as agroecosystems, urban, suburban areas, transport corridors, etc.) and 'pristine areas' dominated by native biota and still experiencing nominal external forcings (c.f. Williams, 1997). Adventive systems have their own ecosystem properties. While minor changes in biotic composition and biogeochemical functioning might be viewed as trivial, the argument that pristine ecosystems still exist is largely untenable. Quantifying the distance that an ecosystem has travelled from its historical range provides a 'reality check' of sorts to those interested in restoration. Further, the difference in ecosystem services provided by the adventive ecosystem provides a metric to assess sustainability concerns.

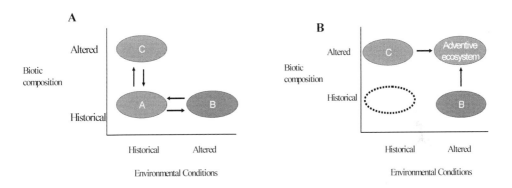

Figure 2 Creation of adventive ecosystems via biotic or abiotic change (modified from Suding *et al.*, 2004). A. an ecosystem is altered by environmental drivers (A→B) or the addition of an abundant, invasive species (A→C). Once in the new state, internal restructuring due to new competitive relationships further alters community composition (via species losses) and through changes in biogeochemical interactions. Restoration of such systems is problematic, if not impossible.

Adventive ecosystems can be constructed *via* two mechanisms. Ecosystems can be removed from their historical trajectories via physical drivers, or ecosystems can be altered by biological additions or deletions (Figure 2). With changes in biotic composition, measurable change in at least a subset of ecosystem processes is likely (Ehrenfeld, 2003). Once a system has been removed from its historical range of variation in terms of climate, atmospheric chemistry, or disturbance return cycles, species changes and subsequent changes in biogeochemical cycling are almost certain. The causal mechanisms for change can be of biotic or abiotic origin, but the outcome may be very similar. Because both biota and environmental drivers have been altered, returning the system to its nominal state is hypothesised to be a low probability event without the application of extraordinary management procedures.

Enhanced Nitrogen Deposition + Invasive Species = Adventive Grasslands

The above 'formula' is suggested to be the current nominal status of grasslands in the USA and perhaps many other regions of the world as well. Nitrogen availability has increased above historical levels in many and perhaps most grassland ecosystems. The global source is atmospheric N deposition (Vitousek *et al.* 1997; Townsend *et al.* 2003), but this is often accompanied in grasslands by enrichment nutrient resulting from fire suppression (Blair, 1998) and by chronic grazing (McNaughton *et al.*, 2001; Johnson & Matchett, 2001). Increased N availability has a high probability of:

1. Reducing plant species richness and diversity (Stevens *et al.*, 2004).
2. Favouring a change in the functional composition of selected grasslands from C_4 dominance to C_3 dominance (especially in conjunction with fire suppression; Knapp *et al.*, 1998; Collins *et al.*, 1998).
3. Directly or indirectly reducing mycorrhizal activity (Egerton-Warburton *et al.*, 2001).
4. Favouring a subset of soil decomposers, potentially those that may stimulate decomposition and mineralisation (Callaham *et al.*, 2003; Madritch & Hunter, 2003, De Deyn *et al.*, 2004; Wall *et al.*, unpublished results), but may also be reducing the decay rates of more recalcitrant organic fractions (Neff *et al.*, 2003).
5. Slowing or inhibiting restoration efforts aimed at maintaining the historical composition of grasslands (Baer *et al.*, 2003).

The above list, not intended to be inclusive, provides recent examples of phenomena that have been under investigation for some time, and all results tend to expand upon or be supported by an abundant literature that has accumulated in the last decade. The cumulative findings of these studies suggest that deposition rates are increasing or having cumulative effects, and no grassland system is totally immune from these changes. The propensity for large-scale 'eutrophication' of plant communities by various N sources was also noted by Grime (2001). Enhanced atmospheric CO_2 may negate some of the effects of N enrichment (Hu *et al.*, 2001), but global warming has the potential to again favour the enrichment effect via enhanced decomposition and mineralisation. The lack of woody storage substrates and relatively low C:N ratio of grassland soils argue that, in contrast to forests, significant transient increases of C and N storage in grassland soils in response to enhanced N deposition are, in general, unlikely.

Increased N facilitates plant invasions in many cases (Hobbs & Huenneke, 1992). The exceptions in grasslands appear limited to those areas where invaders also facilitate more frequent fire return intervals (c.f. Ehrenfeld, 2003). Nitrogen deposition also appears to be facilitating the spread of invasive earthworm species (Callaham *et al.*, 2003). This earthworm invasion in North American grasslands may have very large effects on carbon storage and N flux since pre-existing soil faunas were often depauperate or even lacking such 'keystone' species.

Will invasions, alone, cause reductions in soil biodiversity? At present, available data exist only for plants, and these indicate that species introductions are exceeding species extirpations at regional scales (Hobbs & Mooney, 1998). A reasonable (and testable) hypothesis is that plant species changes will be responsible for much of the variation in soil communities of grasslands, and this trend could enhance soil biological diversity. In reality, the question is moot. Our research focus needs to be upon the interaction of current drivers affecting soil biodiversity, not the individual effects of the drivers, themselves. Enhanced N deposition, in conjunction with habitat destruction, fragmentation, and other drivers that amplify N enrichment effects, will 'eutrophy' these systems, reduce resource heterogeneity and in all likelihood over-ride any potential increases in soil biotic diversity associated with species invasions.

The recognition of our changing grasslands as adventive ecosystems offers a modest paradigm shift for soil biologists. Historically, the decomposer food web was viewed as a bottleneck for nutrients, especially N, and soil fauna were viewed as essential for providing nutrients for plant productivity (e.g. Crossley, 1977; Seastedt, 2000). Now, both basic and applied ecologists are concerned that these organisms or their newer counterparts in adventive ecosystems will contribute to excess nutrient export from managed and natural ecosystems (e.g. Bohlen et al., 2004). Nitrogen retention and transformation to biologically inert forms historically has been an extremely important service provided by many if not most grasslands. Proactive management techniques to reverse the trend of N pollution in the form of nitrous oxide to the atmosphere and nitrate to aquatic systems should be a logical component of conservation management programmes for these ecosystems. In semiarid grasslands, for example, frequent fire can at least slow if not prevent changes associated with N enrichment (Seastedt et al., 1991; Knapp et al., 1998; Collins et al., 1998), and such management maximises a key ecosystem service, plant productivity.

Examination and manipulation of individual biotic and abiotic variables within the soil independent of other concurrent changes may be a productive academic exercise, but such experiments should be conducted within a broader framework to have value to management. Adventive ecosystems are created by synergistic interactions among drivers and biota, not forcing functions acting independently. The advice to managers therefore remains unchanged by that provided by Leopold (1949) and recently by Hooper et al. (2005): do not throw away the parts while we continue to diagnose the system. Reducing the uncertainties associated with global environmental change and the emergence of adventive ecosystems requires realistic, complex experiments, in conjunction with expanded inventory and monitoring activities. While this conclusion is not new, the acknowledgement that adventive ecosystems will require proactive management activities to maximise conservation and ecosystem service values cannot be understated.

Acknowledgements

I thank Dr. Carl Bock for suggesting the term 'adventive ecosystems' as one that best describes these altered ecosystems. Drs. John Blair and Heather Reed helped improve earlier versions of this manuscript. Dr. Diana Wall's group at CSU did the brunt of the Konza work. Our soil biodiversity research was supported by a National Science Foundation grant, DEB 9806438, to the University of Colorado.

References

Andre, H. M., X. Ducarme & P. Lebrun (2001). Soil biodiversity: Myth, reality or conning? *Oikos,* 96, 3-24.

Baer, S. G., J. M. Blair, S. L. Collins & A. K. Knapp (2003). Soil resources regulate productivity and diversity in newly established tallgrass prairie. *Ecology,* 84, 724-735.

Baker, H. G. 1978. Invasion and replacement in Californian and neotropical grasslands. *In:* J. R. Wilson(ed.) *Plant Relations in Pastures.* CSIRO, Melbourne Australia, 368-383.

Bardgett, R. D., & D. A. Wardle (2003). Herbivore-mediated linkages between aboveground and belowground communities. *Ecology,* 84, 2258-2268.

Blair, J. M., T. C. Todd & M. A. Callaham, Jr. (2000). Responses of grassland soil invertebrates to natural and anthropogenic disturbances. *In:* D. C. Coleman and P. F. Hendrix (eds.) *Invertebrates as Webmasters in Ecosystems,* CAB International Press. pp 43-71.

Bohlen, P. J., S. Scheu, C. M. Hale, M. A. McLean, S. Migge, P. M. Groffman & D. Parkinson (2004). Invasive earthworms as agents of change in north temperate forests. *Frontiers in Ecology and the Environment,* 2, 427-435

Coleman, D. C., D. A. Crossley, Jr., P. F. Hendrix (2004). *Fundamentals of Soil Ecology.* Elsevier Academic, Burlington, MA.

Collins, S. L., A. K. Knapp, J. M. Briggs, J. M. Blair & E. M. Steinauer (1998). Modulation of diversity by grazing and mowing in native tallgrass prairie. *Science,* 280, 745-747.

Crossley, D. A., Jr. (1977). The roles of terrestrial saprophagous arthropods in forest soils: Current status of concepts. *In:* Mattson, W. J. (ed). *The Role of Arthropods in Forest Ecosystems.* Springer-Verlag, New York, pp 49-56.

Callaham, M. A., J. M. Blair, T. C. Todd, D. J. Kitchen & M. R. Whiles (2003). Macroinvertebrates in North American tallgrass prairie soils: effects of fire, mowing, and fertilization on density and biomass. *Soil Biology and Biogeochemistry,* 35, 1079-1093.

Callaway, R. M., G. C. Thelen, A. Rodriguez & W. E. Holben (2004). Soil biota and exotic plant invasion. *Nature,* 427, 731-733.

De Deyn, G. B., C. E. Raaljmakers, H. R. Zoomer, M. P. Berg, P. C. de Rulter, H. A. Verhoef, T. M. Bezemer, and W. H. van der Putten (2003). Soil invertebrate fauna enhances grassland succession and diversity. *Nature,* 422, 711-713.

De Deyn, G. B., C. E. Raaijmakers, J. van Ruijven, F. Berendse & W. H. van der Putten (2004). Plant species identity and diversity effects on different trophic levels of nematodes in the soil food web. *Oikos,* 106, 576-586.

Egerton-Warburton, L. M., R. C. Graham, E B. Allen & M .F. Allen (2001). Reconstruction of the historical changes in mycorrhizal fungal communities under anthropogenic nitrogen deposition. *Proceedings of the Royal Society of London Series B-Biological Sciences,* 268, 2479-2484.

Ehrenfeld, J. G. (2003). Effects of exotic plant invasions on soil nutrient cycling processes. *Ecosystems,* 6, 503-523.

Grime, J. P. (2001). *Plant Strategies, Vegetation Processes, and Ecosystem Properties* (Second Edn). John Wiley & Sons, Chichester.

Hobbs, R. J. & L. F. Huenneke (1992). Disturbance, diversity, and invasion: Implications for conservation. *Conservation Biology,* 6, 324-337.

Hobbs, R. J. & H .A. Mooney (1998). Broadening the extinction debate: Population deletions and additions in California and Western Australia. *Conservation Biology,* 12, , 271-283.

Hooper, D. U., F. S. Chapin, III, J. J. Ewel, A. Hector, P. Inchausti, S. Lavorel, J. H. Lawton, D M. Lodge, M. Loreau, S. Naeem, B. Schmid, H. Setala, A. J. Symstad, J. Vandermeer, & D. A. Wardle (2005). Effects of biodiversity on ecosystem functioning: A consensus of current knowledge. *Ecological Monographs* (in press).

Hu, S., F. S. Chapin, M. K. Firestone, C. B. Field & N. R. Chiariello (2001). Nitrogen limitation of microbial decomposition in a grassland under elevated CO_2. *Nature,* 409, 188-191.

Johnson, L. C. & J. R. Matchett (2001). Fire and grazing regulate belowground processes in tallgrass prairie. *Ecology,* 82, 3377-3389.

Klironomos, J. N (2003). Variation in plant responses to native and exotic arbuscular mycorrhizal fungi. *Ecology,* 84, 2292-2301.

Knapp, A. K., J. M. Briggs, D. C. Hartnett & S. L. Collins (eds.) (1998) *Grassland Dynamics: Long-Term Ecological Research in Tallgrass Prairie.* Oxford University Press. New York.

Landres, P. B., P. Morgan & F. J. Swanson 1999. Overview of the use of natural variability concepts in managing ecological systems. *Ecological Applications,* 9, 1179-1188.

Leopold, A. (1949). *A Sand County Almanac.* Oxford University Press. New York.

Madritch, M. D. & M. D. Hunter (2003). Intraspecific litter diversity and nitrogen deposition affect nutrient dynamics and soil respiration. *Oecologia,* 136, 124-128.

McNaughton, S. J., F. F. Banyikwa & M. M. McNaughton (2001). Promotion of the cycling of diet-enhancing nutrients by African grazers. *Science,* 278, 1798-1800.

Mikola, J., R. D. Bardgett & K. Hedlund (2002). Biodiversity, ecosystem functioning and soil decomposer food webs. *In*: M. Loreau, S. Naeem and P. Inchausti. (eds). *Biodiversity and Ecosystem Functioning: Synthesis and Perspectives.* Oxford University Press, pp 169-180.

Moore, J. C., K. McCann, H. Setala & P. C. De Ruiter (2003). Top-down is bottom-up: Does predation in the rhizosphere regulate aboveground dynamics? *Ecology,* 84,, 846-857.

Moore, J. C., E. L. Berlow, D. C. Coleman, P. C. de Ruiter, Q. Dong, A. Hastings, N. C, Johnson, K. S. McCann, K. Melville, P. J. Morin, K. Nadelhoffer, A. D. Rosemond, D. M. Post, J. L. Sabo, K. M. Scow, M. J. Vanni, & D. H. Wall (2004). Detritus, trophic dynamics and biodiversity. *Ecology Letters,* 7, 584-600.

Neff, J. C., A. R. Townsend, G. Gleixner, S. J. Lehman, J. Turnbull & W. D. Bowman (2002). Variable effects of nitrogen additions on the stability and turnover of organic carbon. *Nature,* 419, 915-917.

Porazinska, D. L., R. D. Bardgett, M. B. Blaauw, H. W. Hunt, A. N. Parsons, T. R. Seastedt & D.H. Wall (2003). Relationships at the aboveground-belowground interface: plants, soil microflora and microfauna, and soil processes. *Ecological Monographs,* 73, 377-395.

Reed, H. E. (2005). Effects of fire and plant invasion on aspects of aboveground and belowground interactions in an eastern tallgrass prairie. Ph.D. Dissertation, University of Colorado, Boulder.

Rice, C. W., T. C. Todd, J. M. Blair, T. R. Seastedt, R. A. Ramundo, & G. W. T. Wilson (1998). Belowground biology and processes. *In*: A. K. Knapp, J. M. Briggs, D. C. Hartnett & S. L. Collins (eds.) *Grassland Dynamics: Long-Term Ecological Research in Tallgrass Prairie.* Oxford University Press. New York, 244-264.

Rudgers, J. A., M. Koslow & K. Clay (2004). Endophytic fungi alter relationships between diversity and ecosystem properties. *Ecology Letters,* 7, 42-56.

Seastedt, T. R. (2000). Soil Fauna and controls of carbon dynamics: Comparisons of rangelands and forests across latitudinal gradients. *In*: D.C. Coleman and P. Hendrix (eds). *Invertebrates as Webmasters of Ecosystems.* CABI Publishing, Wallingford UK.

Seastedt, T.R. (2002). Base camps of the Rockies: The intermountain grasslands. In: J. Baron. (ed). *Rocky Mountain Futures: An ecological Perspective.* Island Press, Covelo CA. pp 219-213.

Seastedt, T. R., J. M. Briggs & D. J. Gibson (1991). Controls of nitrogen limitation in tallgrass prairie. *Oecologia,* 87, 72-79.

Stevens, C. J., N. B. Dise, J. O. Mountford & D. J. Gowing (2004). Impact of nitrogen deposition on the species richness of grasslands. *Science,* 303, 1876-1879.

Suding, K. N, K. L. Gross & G. R. Houseman (2004). Alternative states and positive feedbacks in restoration ecology. *Trends in Ecology and Evolution,* 19, 46-53.

Todd, T. C. (1996). Effects of management practices on nematode community structure in tallgrass prairie. *Applied Soil Ecology,* 3, 235-246.

Townsend, A. R., R. W. Howarth , F. A. Bazzaz , M. S. Booth , C. C. Cleveland , S. K. Collinge , A. P. Dobson , P. R. Epstein, E. A. Holland , D. R. Keeney , M. A. Mallin , P. Wayne & A. Wolfe (2003). Human health effects of a changing global nitrogen cycle. *Frontiers in Ecology and the Environment,* 1, 240-246.

Van der Heijden, M.G.A., J.N. Klironomos, M. Ursic, P. Moutoglis, R. Streitwolf-Engel, T. Boller, A. Wiemken & I.R. Sanders (1998). Mycorrhizal fungal diversity determines plant biodiversity, ecosystem variability and productivity. *Nature,* 396, 69-72.

Vitousek, P. M., J. D. Aber, R. W. Howarth, G. E. Likens, P. A. Matson, D. W. Schindler, W. H. Schlesinger, & D. G. Tilman (1997). Human alteration of the global nitrogen cycle: Sources and consequences. *Ecological Applications,* 7, 737-750.

Wardle, D. A. (2005). How plant communities influence decomposer communities. *In*: R. D. Bardgett, D. W. Hopkins & M. B. Usher. (eds) *Soil Biodiversity and Ecosystem Functioning..* Cambridge University Press.

Wardle, D. A. (2002). *Communities and Ecosystems: linking the aboveground and belowground components.* Princeton University Press.

Wardle, D. A. & W. H. van der Putten. (2002). Biodiversity, ecosystem functioning and above-ground - below-ground linkages. *In*: M. Loreau, S. Naeem, and P. Inchausti. (eds). *Biodiversity and Ecosystem Functioning: Synthesis and Perspectives.* Oxford University Press, pp 155-168.

Wardle, D. A., R. D. Bardgett, J. N. Klironomos, H. Setala, W. H. van der Putten, & D. H. Wall (2004). Ecological linkages between aboveground and belowground biota. *Science,* 304, 1629-1633.

Williams, C. E. (1997). Potential valuable ecological functions of nonindigenous plants. *In*: J. O. Luken & J. W. Thieret (eds). *Assessment and Management of Plant Invasions.* Springer, New York, pp 26-34.

Chemical components and effects on soil quality in temperate grazed pasture systems

M.H. Beare, D. Curtin, S. Thomas, P.M. Fraser and G.S. Francis
New Zealand Institute for Crop & Food Research, Canterbury Agriculture and Science Centre, Private Bag 4704, Christchurch, New Zealand, Email: Bearem@crop.cri.nz

Key points

1. Legume/grass pastures retain feedbacks on N supply that may help to reduce losses of N to the wider environment.
2. Intensive use of N fertilisers tends to increase SOM turnover and increase losses of N to the environment.
3. Increased use of fertilisers has contributed little to soil organic matter storage in grazed pastoral systems.

Keywords: soil quality, grassland, pasture, chemical fertility, compaction, GHG emissions

Introduction

Soil quality is commonly defined as the *fitness of soil for a specific use* (Larson & Pierce, 1994). Much of the interest in soil quality pertains to the management of agricultural systems, i.e. the soil's ability to nurture and sustain plant and animal productivity (Beare, 2002), partition and regulate the flow of water, and function as an environmental buffer (Carter *et al.*, 1997; Doran & Parkin, 1994). Chemical, physical and biological indicators of soil quality reflect the key properties and processes that support these functions and are aimed at assessing the soil's ability to satisfactorily provide them. In addition to extrinsic factors (e.g. climate), these functions are influenced both by the intrinsic characteristics of a soil (i.e. inherent soil quality) and the properties and processes that are influenced by its use and management (i.e. dynamic soil quality). The properties of greatest importance to soil quality in a particular agricultural system are often grouped into a minimum data set (Gregorich *et al.*, 1994). These are best defined by the soil management issues (e.g. nutrient availability, compaction, water storage) that have the most influence on plant and animal performance and impact on the wider environment (Larson & Pierce, 1994; Beare *et al.*, 1999). Sustainable production depends on choosing land uses that are well suited to the capability of the soil (and wider environment) and maintaining soil conditions that enhance productivity and environmental quality (Larson & Pierce, 1994; Beare, 2002).

The term 'grassland' covers a wide range of managed and unmanaged ecosystems that differ in their productivity, nutrient cycling, and pathways of herbage removal and return. Grasslands managed for livestock grazing are common in many regions of the world. They include legume/grass and grass-based pastures ranging from extensive low input rain-fed sheep systems to intensive high input irrigated dairy systems. This paper addresses the chemical components of dynamic soil quality in managed grasslands, with a particular focus on temperate grazed pasture systems. It explores how chemical inputs, both natural and anthropogenic, affect the productivity and environmental quality of these pastoral soils. It also describes how management influences chemical properties and processes through effects on the physical and biological condition of grazed pastures, which in turn impact on the sustainable management of these grasslands.

Chemical inputs and fertility

Soil chemical fertility testing represents one of the earliest and best established approaches to soil quality monitoring. Although chemical fertility describes only one aspect of soil quality, it remains an essential component of any comprehensive soil quality assessment programme. Fertility testing of grassland soil usually involves assessment of pH, plant available phosphorus (P), exchangeable cations (K^+, Ca^+, Mg^{2+}, Na^+), and SO_4^{2-} sulphur. Apart from mineral nitrogen (N) (NH_4^+, NO_3^-), widely accepted, commercially available tests for plant available N are decidedly lacking. Micronutrient (e.g. Mn, Cu, Mo, B, Se) testing is not normally included in regular monitoring programmes; rather it is usually directed at identifying specific nutrient deficiencies. Limitations of these tests and implications for soil quality monitoring are discussed below.

Efforts to improve the productivity of grassland systems have relied heavily on the use of fertilisers. Legumes are an important component of grazed pastoral systems in Australia, New Zealand, Western Europe and both North and South America (Peoples et al., 2004). In Australia and New Zealand, P and sulphur (S) have traditionally been the main nutrients applied to these pastures, usually in the form of single superphosphate. The benefits of superphosphate for herbage dry matter (DM) production are well established (Haynes & Williams, 1993). Ensuring high P availability is especially important in stimulating clover growth and symbiotic N_2-fixation, which can markedly improve the N status of legume-based pastures (Lambert et al., 2000; Ledgard, 2001). Despite the obvious benefits of superphosphate for pasture and animal productivity, it may also have important adverse effects on soil and environmental quality. The increased use of P fertilisers and importation of P in feed, manure and effluent has resulted in accumulation of soil P in many locations worldwide (Haygarth & Jarvis, 1999; Sims et al., 2000). Haygarth et al. (1998) reported annual accumulation rates of 0.28 and 26 kg/ha for extensive sheep and intensive dairy pastures in the UK, respectively. There is mounting evidence that P transfer from agricultural land to water bodies also increases with adverse impacts on water quality (e.g. Daniel et al., 1998). High levels of P accumulation in soils are at least partly responsible for this increased transfer. The processes and pathways involved in the transfer of diffuse P from soil to surface and subsurface waters have been described in a number of recent reviews (e.g. Haygarth & Jarvis, 1999) and involve both soil solution P and mineral bound P (McDowell et al., 2001).

Relationships between P transport in overland or subsurface flow and soil P levels have been investigated extensively, prompted by a desire to establish management thresholds to assist in mitigation of P transport from agricultural landscapes (Sims et al., 2000). Several studies have shown markedly increased rates of P transfer when plant available P (e.g. Olsen P) exceeds a critical level (e.g. McDowell et al., 2001; Maquire & Sims, 2002). This critical level is often termed the 'change point' and is closely related to the degree of soil P saturation. Change point data have been used to assess risks of P transfer in subsurface and overland flow (McDowell & Sharpley, 2001). Soil P status is only one of the factors contributing to losses from pastoral soils. Transport of P from the landscape requires conditions favourable to overland or subsurface flow. Slope and drainage strongly influence the pathways of transfer at a field scale. The latter is affected by soil type and texture, structural properties (e.g. macroporosity) influencing infiltration, and plant cover. Preferential flow through root channels and earthworm burrows may help to bypass sorption surfaces and thereby promote rapid transfer of P down the soil profile (Haygarth & Jarvis, 1999). The magnitude of losses are also affected by the size of rainfall events and amount and timing of fertiliser applications.

Despite the advantages of legumes for improving soil N status and forage quality, there are also a number of disadvantages. In comparison to grasses, legumes tend to compete less effectively for nutrients, grow more slowly and produce less dry matter. They also tend to be more susceptible to cold, wet conditions; leading to greater seasonal variation in their DM production. Because of these problems, intensification of pastoral farming has led to the sowing of all-grass pastures and an increased reliance on fertiliser N in place of legume fixed N. In New Zealand, for example, urea use increased from 50,000 t/yr in 1992 to 310,000 t/yr in 2002. Between 1996 and 2002, fertiliser N rates increased from 39 to 102 kg/ha on dairy farms as compared with an increase from 0.7 to 6 kg/ha on sheep/beef farms. For each kg of fertiliser N applied, biologically fixed N may decrease by 0.3 to 0.7 kg/ha (Saggar, 2004). It is well known that the use of N fertilisers reduces the clover content and N-fixation rates of grass-clover swards (Ledgard et al., 2001).

Soil pH tends to decline with time under improved pasture management. The build up of surface soil organic matter (SOM) and N fertility in legume-based pastures results in a gradual increase in cation exchange capacity and H^+ saturation through high cation uptake by N_2-fixing legumes. The decline in pH is a consequence of high rates of N-fixation, greater mineralisation of organic matter (nitrification in urine patches) and leaching of nutrients down the profile (Haynes & Williams 1993). Relatively small decreases in soil pH can lead to large reductions in pasture production because of a build-up of phytotoxic levels of Al and Mn. Soil acidity is a problem that can be readily managed by regular monitoring of soil pH and appropriate applications of lime. Earthworm populations tend to increase following additions of lime to acidic soils (Springett, 1985) which may lead to increased organic breakdown and faster cycling of N and P as discussed below. Care should be taken to avoid excessive liming, which can change the availability of trace elements (e.g. iron, manganese, copper, cobalt) and increase the risk of some soil-borne diseases.

Applications of sewage sludge to grasslands is increasing in many areas of the world (Towers & Horne, 1997). While sewage sludge can have significant agronomic benefits as a source of nutrients (N, P) and organic matter, it can also increase soil greenhouse gas (e.g. nitrous oxide - N_2O) emissions (Scott et al., 2000) and the eutrophication of water bodies (Towers & Horne, 1997). Although, the application of sewage sludge can contribute to OM accumulation and C sequestration, it is less effective in this regard than other land management practices (Smith et al., 2000), particularly in grassland systems. The adverse impacts of sewage sludge are most often associated with the inputs of heavy metals such as Cd, Zn, Cu, Pb and Ni (Towers & Horne, 1997). Heavy metals (principally Cd, Zn and Ni) enter the food chain through plant uptake, which tends to be greatest in acidic and coarse textured soils (Hooda et al., 1997). Many grassland soils have a relatively strong metal binding capacity at neutral to basic pH levels. Soil acidification could markedly reduce this binding capacity and, therefore, increase the risk that heavy metals will enter the food chain and pollute water. This information will be important for interpreting soil pH data to improve the management of soil quality on sewage sludge amended soils.

Soil organic matter and soil quality

SOM contributes to soil quality through its effects on a wide range of soil properties and processes that influence soil physical condition and fertility. Grassland soils can have substantially more SOM than comparable soils used for arable cropping because of continuous inputs of plant roots, senescent herbage and animal dung. SOM content represents the balance between inputs and decomposition of organic matter. Although nutrient (N, P) application can substantially increase the amount of organic matter returned to the soil in plant litter and dung, studies suggest that there is relatively little effect on SOM in the long term (Metherell, 2003).

The Winchmore trial in New Zealand has provided long-term data on the effects of superphosphate on SOM in an irrigated ryegrass-clover pasture used for sheep grazing. The trial, which consists of three rates of superphosphate (0, 188, and 376 kg/ha), was initiated in 1952 on an intensively cropped, degraded soil. Superphosphate increased herbage dry matter and also the proportion of clover in the pasture. Annual average DM production (1952-2002) was 4.8 t/ha in the control, increasing to 11.9 t/ha at the 376 kg/ha superphosphate rate (Table 1). SOM increased rapidly during the first 15 years following pasture establishment. Contrary to the common perception that SOM increases with increasing DM production (and thus increased organic matter inputs), recent measurements of soil C show no significant increase associated with fertiliser addition. A similar steady-state soil C level was achieved, regardless of whether or not superphosphate was used. Tate et al. (1997) also concluded that there was no evidence for a net accumulation of soil C (i.e. C sequestration) under long-term pasture management based on monitoring of soil C levels from a large number of New Zealand grassland sites over the last 50 years, despite large increases in fertiliser-derived production. Lack of a SOM response to this very large increase in DM production is consistent with suggestions that above-ground C inputs have less influence on soil C storage than root C. Recently, Rasse et al. (2004) estimated that the mean residence time of root C in soil is 2.4 times that of shoot C. Evidence from the Winchmore trial shows that plants growing in low P soil have larger root systems that turnover more slowly than those in P fertilised soil (Table 1). Where native (e.g. tussock) grasslands have been oversown by improved grass/clover mixes, some studies suggest that there may be an inverse relationship between pasture production (and associated differences in stocking rate) and soil C content (Metherell, 2003).

Table 1 Effects of long-term superphosphate rates on above and below ground production, and turnover and soil organic C levels (adapted from Metherell, 2003)

	Superphosphate rate (kg/ha per year)			LSD
	0	188	376	p=0.05
Herbage production (t/ha per year)	4.8	10.9	11.9	0.9
Root mass (t C/ha per year)	2.8	-	2.1	0.2
Root turnover time (years)	1.9	-	1.3	0.7
Root production (t C/ha per year)	1.7	-	1.7	Ns
Total soil C (t C/ha)	72.3	75.1	73.0	Ns

Clover requires relatively high soil P levels and fertilisation is important to stimulate clover growth and symbiotic fixation of N. Superphosphate can substantially increase the quantities of N cycling in legume-based pasture (Metherell, 2003) and some of this may be retained in SOM. Increases in total soil N following phosphate application have been reported by Haynes & Williams (1992) and Parfitt et al. (2005). However, annual increases in soil N under high v. low P fertility pasture may not be large (e.g. 19 kg/ha per annum; Lambert et al., 2000): once steady state is achieved, further retention of N is likely to be low. Changes in SOM quality in response to fertilisation include decreased ratios of C-to-N, C-to-organic P, and C-to-S (Haynes & Williams, 1992), and increased microbial biomass and mineralisable N (Parfitt et al., 2005). These changes should generally result in more rapid cycling of nutrients. Results from Parfitt et al. (2005) indicate that N mineralisation in clover-ryegrass pasture increases as soil P status (Olsen P test) increases and Haynes and Williams (1992) showed higher rates of S mineralisation in soil fertilised with superphosphate compared with unfertilised soil.

Biological indicators and impacts

Microorganisms and soil fauna contribute in many different ways to the chemical fertility of grassland soils. Their populations and activity are also affected by chemical inputs to soils. Soil fertility has been shown to influence both the size and turnover rate of biologically active pools of soil organic matter. There have been numerous reports of improvements in soil N status under high fertility legume-based pastures (Lamberts et al., 2000; Ledgard, 2001) with annual increments in total N commonly ranging from 30 to 80 kg N ha[-1] (Peoples & Baldock, 2001). Most studies report that active pools of organic matter show a positive to neutral response to fertiliser P (Tate et al., 1991; Sarathchandra et al., 1994; Ghani et al., 2003). Ghani et al. (2003) provided results from several fertility studies that showed increasing levels of microbial biomass C and N and hot water extractable C with increases in P fertility. This effect appears to be related to differences in above ground dry matter production along P fertility gradients.

In contrast to P, applications of N to established pastures often have adverse effects on active pools of organic matter (Okano et al., 1991; Ghani et al., 2003) that may be attributed to an increase in shoot-to-root ratios, a reduction in root biomass and increased turnover of biologically active organic matter pools. Ghani et al. (2003) showed that microbial biomass C and N, and hot water extractable C decreased with increases in N fertiliser applications from 0 to 400 kg/ha. Results from their monitoring of dairy (n=12) and sheep/beef (n=12) farms (Allophanic soils) showed that clover/grass pastures fertilised with P and grazed by sheep had significantly higher levels of microbial C, N and S, mineralisable N and hot water extractable C than intensively managed, N fertilised dairy pastures. Ghani et al. (2003) recommended hot water extractable C (HWC) as a useful indicator of soil quality based on its strong correlation to other commonly used measures of organic matter content/quality (i.e. microbial biomass C and N, mineralisable N and total carbohydrates) and its sensitivity to grassland fertiliser management practices. These authors also showed that there can be fairly significant inter-annual (seasonal) and intra-annual variation in these indicators. Understanding the effects of this variation on OM turnover and nutrient cycling is important in providing practical interpretations of this measurement.

European earthworms are now commonly found in temperate pastoral soils around the world. In New Zealand, their introduction to grazed pastures has been shown to lift pasture production by an average of 25-30% or the equivalent of about 2.5 stock units/ha (Stockdill, 1982). Earthworms perform a number of important functions in soils that influence plant nutrition and pasture production. They have been shown to increase the solubility of plant nutrients (Sharpley & Syers, 1976), accelerate mineralisation of organic N (Willems et al., 1996) and assist in the incorporation of surface-applied lime in pastures (Springett, 1983). On hill country pastures, reductions in runoff and soil erosion have been attributed to improvements in infiltration resulting from earthworm activity. However, other studies suggest that earthworms may lower the quality of runoff water (e.g. Sharpley & Syers, 1976). Their effects on soil aggregation and macropore formation can also contribute to increased bypass flow and greater leaching of surface-applied pesticides and nitrogen (Edwards et al., 1992). Because much of the existing data pertains to cropping soils, there is a need for more research on the benefits and impacts of earthworms on the chemical quality of grassland systems, in particular their impact on the accumulation and loss of SOM.

Nutrient cycling

The effects of livestock farming practices on nutrient cycling depend on whether the nutrients removed in forage are returned as dung and urine during grazing, as effluent, as fertilisers, or as a combination of these. The amount of nutrients returned to soil from livestock excreta differ widely between pastoral farming systems. Dairy cows, for example, excrete ca 60-99% of the nutrients that they ingest from forage and feed. Some nutrients (e.g. P, Ca, Mg, Cu, Zn etc.) are excreted predominantly in faeces while others (e.g. K) are excreted mainly in urine (Haynes & Williams, 1993). Significant amounts of N, Na, Cl, and S are usually found in both urine and faeces. A relatively high proportion of the ingested N (21%), P (35%) and Ca (22%), but a low proportion of the K (7%) is removed in meat and milk (Haynes & Williams, 1993). The urine patches of dairy cows can have very high loadings of N (1000 kg/ha), K (500 kg/ha) and other nutrients (e.g. Ca, Mg, S), often exceeding the amount that pasture can take up during a growing season (Williams et al., 1990; Haynes & Williams, 1993; Di & Cameron, 2002). Once mineralised, the surplus nutrients are susceptible to leaching (i.e. NO_3^-, K^+, Ca^{2+}, Mg^{2+}) where there is adequate drainage. Following the hydrolysis of urea to NH_4^+, nitrification and denitrification can contribute to large emissions losses of N as N_2 and N_2O.

Nitrogen fixation in legume-based pastures is regulated by soil inorganic N levels and pasture species composition. Increases in soil available N tend to lower the legume component of pastures and, as a consequence, decrease the N derived from fixation. This regulation of N_2 fixation acts to limit the N input from legumes in legume/grass pastures and thereby reduce the potential for N losses to the wider environment (Ledgard, 2001; Figure 1). Ledgard (2001) compiled data from several studies which showed that NO_3^- leaching losses are similar for grazed legume/grass and N-fertilised grass pastures at similar rates of N input. Furthermore, available data suggest that the proportion of total N lost (to volatilisation, denitrification and NO_3 leaching) to the wider environment tends to increase as the rate of N input increases.

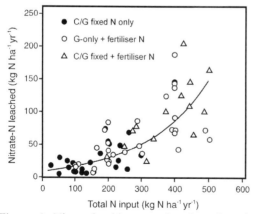

Figure 1 Nitrate leaching as a function of total N inputs from clover/grass (C/G) or grass only pastures, with or without N fertiliser additions (redrawn from Ledgard, 2001)

Nitrate leaching from grasslands is a major environmental concern in many countries (Jarvis et al., 1995). This is a particular concern under intensive land uses where there are high inputs of N in the form of fertilisers or effluents, or where N is returned in the urine of grazing animals. Goh & Williams (1999) compared N budgets for a wide range of temperate grazed pastoral systems. Their results showed that losses of N by leaching, volatilisation and denitrification from clover-based sheep/beef and dairy pastures were relatively low (7 - 74 kg/ha per year) and much less than the N removed in animal products. Nitrogen budgets for extensive, clover/grass sheep production systems in New Zealand and the UK indicate that leaching losses are very low (5-14 kg/ha per year) and much less than the total N inputs. By comparison, intensive N fertilised dairy and beef production systems in the Netherlands and the UK have reported N losses up to 390 kg/ha per year. In intensive systems, leaching can account for >50% of the total N losses and represent a high proportion of the total N inputs. In New Zealand, N_2O accounts for *ca* 20% of the country's greenhouse gas inventory, 50% of which has been attributed to animal excreta (de Klein *et al.*, 2001). Recently, Di & Cameron (2002) showed that treating urine patches with a nitrification inhibitor (dicyandiamide [DCD]) can reduce nitrate leaching from these areas by 42-76%. Leaching of cations (i.e. K^+, Ca^{2+}, Mg^{2+}) that typically accompany NO_3^- was also reduced by 50-65% where DCD was applied (Di & Cameron, 2004).

Spatial variability in the chemical characteristics of grazed pastures can be very high. The location of shelter, water sources, gateways and farm roads can cause the establishment of fertility gradients resulting from uneven deposition of livestock dung and urine (Matthew *et al.*, 1994). This effect can be particularly pronounced on hill country sheep farms where, for example, Ledgard (2001) estimated a net loss of 40 kg N/ha year in excreta from steep slope areas and a net gain of 180 kg N/ha year in stock camps on lower slopes. These nutrient gradients in turn can contribute to variation in the species composition and growth rate of pastures which can affect nutrient removal and N_2-fixation (Ledgard, 2001).

Physical effects of livestock treading

The detrimental effects of livestock treading on soil physical properties have been described in several studies (e.g. Drewry *et al.*, 2004, Naeth *et al.*, 1990). Treading can lead to localised, if not wide-spread, compaction of topsoil, especially under wet conditions. Compaction tends to reduce total soil porosity, especially the abundance of macropores. The decrease in macroporosity and associated physical properties has been linked to yield reduction in spring pastures (Drewry *et al.*, 2004, Naeth *et al.*, 1990). Compaction also retards water infiltration and gas diffusivity, leading to the development of anaerobic sites where N_2O production can be high. On average, N_2O emission from grazed pastures is more than twice that of mown pastures. Oenema *et al.* (1997) estimated that emissions from grazed pastures accounted for 0.2- 9.9% of the N excreted by grazing animals.

Intensification of livestock farming on high P fertility soils may increase the risk that damage from stock treading will increase surface runoff. Evidence from New Zealand hill country farms (steep land, slope >20 degrees) indicates that sediment losses are much greater from cattle-grazed (2740 kg/ha per year) than from sheep-grazed (1220 kg/ha per year) pastures (Lambert *et al.*, 1985). In this study, the P fertility level (and associated differences in stocking rate) had no effect on sediment loss, possibly because there was less runoff from the high fertility pastures because of better surface cover. In contrast, McDowell *et al.* (2003) showed that increased losses of sediment and P were associated with reductions in macroporosity resulting from cattle treading. Additional research is needed to quantify the relationships between soil characteristics (e.g. texture, OM content etc), P fertility, grazing type and intensity, and sediment losses. Furthermore, little is known about effects of pastoral management practices on the transport of other sediment-bound compounds (e.g. Cd, DDT).

Intensification of livestock farming has also resulted in an increased reliance on supplementary feed and forage crops, particularly during periods of lower pasture production. In New Zealand and elsewhere, forage crops are often grown as a break-crop prior to resowing pastures. Stock treading on cultivated soils can cause severe compaction which may adversely affect subsequent crop production. Thomas et al. (2004) showed that using no-tillage practices to establish winter forage crops (ex pasture) can reduce soil compaction during grazing under wet conditions ($\geq$ field capacity) and markedly improve the re-growth of multi-graze crops as compared with those established with minimum or intensive tillage practices. The more stable soil surface created by direct drilling crops into long-term pasture appears to lower the risk of surface compaction from heavy stock treading, even under very wet conditions. This reduction in compaction not only improved the re-growth of the forage crops but also greatly reduced N_2O emissions (Table 2: Thomas et al., 2004).

Table 2 Dry matter production and cumulative N_2O flux following simulated treading of crops established with three different tillage practices

Moisture content at treading[1]	Crop re-growth (kg DM/ha)[2]			Cumulative N_2O flux (kg N/ha)		
	IT [3]	MT	NT	IT	MT	NT
< FC	14.3	13.3	15.1	1.45	1.87	2.04
FC	10.9	15.1	14.6	5.73	3.18	3.00
> FC	7.1	9.5	17.8	14.86	12.68	4.97

[1]FC = Field capacity
[2]Triticale crop regrowth (4 months) after herbage removal in June 2003. DM = dry matter
[3]IT = intensive tillage; MT = minimum tillage; NT = no-tillage

Natural amelioration of soil physical properties in compacted grasslands occurs over periods ranging from several months to several years (Drewry et al., 2004). Knowledge of these recovery rates will be important to interpreting the effects of compaction on soil chemical properties and processes and devising monitoring schedules that reflect these trends.

Soil quality monitoring

The use of soil quality information to evaluate the sustainability of soil management systems has involved two approaches (Larson et al., 1994). With the *comparative assessment* approach, the condition of the soil from one system is compared with another using one-off measures of soil properties. This approach is often used to compare the effects of different land uses or management practices. With the *dynamic assessment* approach, the condition of soil under changing land use or management practices is assessed by monitoring soil properties over time. In agricultural systems, applications of the dynamic assessment approach usually involve both monitoring (i.e. repeated measurement) and control (i.e. regulation or management), and the focus shifts from describing soil conditions to maintaining soil conditions that sustain high productivity and minimise environmental impacts through improved management. To this end, soil quality monitoring may be coupled with best management practice recommendations to develop soil management decision support systems that improve the sustainability of production systems. The soil management decision support systems that incorporate soil quality monitoring have been described for mixed-cropping (arable/pastoral) (Beare *et al.*, 1999), extensive sheep/beef (Shepherd, 2000) and dairy farming systems (de Klein *et al.*, 2004). Most of these systems focus heavily on the physical aspects of soil quality and better integration with chemical fertility data is needed.

State-of-the-Environment (SOE) monitoring programmes in New Zealand and elsewhere have shown clear differences in the condition of soils under different grass-based land uses. Sparling & Schipper (2004) recently reported results from a three-year SOE monitoring programme involving a total of 511 sites, representing 11 different land uses, and covering 15 different soil orders. Not surprisingly perhaps, they showed that chemical fertility indicators including available P (i.e. Olsen P), mineralisable N, total C and total N differed among land uses in the order: dairy pastures > drystock (extensive sheep/beef) pastures > tussock grasslands. The higher levels of total C, total N and mineralisable N reflect an increase in SOM content under dairying. The higher Olsen P and mineralisable N values under dairying reflect the regular use of superphosphate fertilisers and resulting improvements in N-fixation.

Regular monitoring of soil fertility status is a key aspect of fertiliser management on many pastoral farms. Soil fertility analyses are usually carried out by private or government testing laboratories, which consequently hold large and potentially useful databases for assessing local, regional and even national trends in soil fertility. Where the economic optimum range and environmental thresholds of different fertility tests have been defined, information of this type can be useful to industry groups and regulatory authorities for sustainable land use planning and policy development. In a recent example from New Zealand, Wheeler *et al.* (2003) evaluated changes in the fertility status (pH, Olsen P, and exchangeable Ca, Mg, and K) of sheep/beef and dairy farms based on analysis of 246,000 commercial samples taken between 1988 and 2001. In terms of soil P status, their analyses showed a general rise in Olsen P values under both pastoral land uses that corresponded with increases in the recommended ranges for optimum economic production (Figure 2). Their data also showed that *ca* 30% of the dairy farms had Olsen P values below the target for maximum dairy production while *ca* 20% could lower fertiliser P inputs without loss of production. Approximately 30% of the sheep/beef farms had Olsen P values below the lower economic target. This coupled with the recent decline in median Olsen P values suggests that there is an opportunity to lift the production and profitability of this sector by increasing soil P levels.

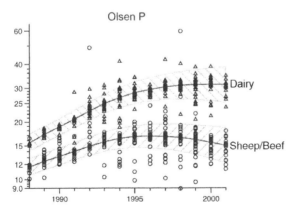

Figure 2 Trends in Olsen-P (log-transformed) in New Zealand sheep/beef and dairy pastures on sedimentary soils (redrawn from Wheeler *et al.*, 2003)

References

Beare, M. H. (2002). Soil quality and productivity. *In:* R. Lal (ed.), *Encyclopedia of Soil Science*, Marcel Dekker, New York, 1078-1081.

Beare, M. H., P. H. Williams & K. C. Cameron (1999). On-farm monitoring of soil quality of sustainable crop production. In: *Best Soil Management Practices for Production*, Fertiliser and Lime Research Centre Conference Proceedings, Palmerston North, New Zealand, 81-90.

Beare, M. H., T. van der Weerden, C. Tregurtha & P. H. Williams (2002). Soil Quality Management System User Manual - for Canterbury arable and mixed cropping farms. Crop & Food Research, Christchurch, 45 p. (ISBN 0-478-10827-3).

Cameron, K. C., H. J. Di & R. G. McLaren (1997). Is soil an appropriate dumping ground for washout wastes? *Australian Journal of Soil Research*, 35, 995-1035.

Carter, M. R., E. G. Gregorich, D. W. Anderson, J. W. Doran, H. H. Janzen & F. J. Pierce. (1997). Concepts of Soil Quality and Their Significance. *In:* E. G. Gregorich & M. R. Carter (eds), *Soil Quality for Crop Production and Ecosystem Health*, Elsevier, Amsterdam, 1-9.

Daniel, T. C., A. N. Sharpley & J. L. Lemunyon (1998). Agricultural phosphorus and eutrophication: a symposium overview. *Journal of Environmental Quality*, 27, 251-257.

De Klein, C., I. Tarbotton, J. Morton, K. Betteridge, M. Beare, S. Officer, C. Tregurtha and B. Thompson (2004). Best management practices for grazing dairy cows: using the Dairy Soil Management System. AgResearch, Palmerston North, New Zealand, 33 p.

De Klein, C. A. M., R. R. Sherlock, K. C. Cameron, T. J. van der Weerden (2001). Nitrous oxide emissions from agricultural soils in New Zealand – a review of current knowledge and directions for future research. *Journal of the Royal Society of New Zealand*, 31, 543-574.

Di, H. J. & K. C. Cameron (2002). The use of a nitrification inhibitor, dicyandiamide (DCD), to decrease nitrate leaching and nitrous oxide emissions in simulated grazed and irrigated grassland. *Soil Use and Management*, 18, 395-403.

Di, H. J. & K. C. Cameron (2004). Effects of the nitrification inhibitor dicyandiamide on potassium, magnesium and calcium leaching in grazed grassland. *Soil Use and Management*, 20, 2-7.

Doran, J. W. & T. B. Parkin (1994). Defining and assessing soil quality. *In:* J. W. Doran, D. C. Coleman, D. F. Bezedick & B. A. Stewart (eds), *Defining Soil Quality for a Sustainable Environment*, Special Pub. No. 35; American Society Agronomy: Madison, 3-21.

Drewry, J. J., R. P. Littlejohn, R. J. Paton, P. L. Singleton, R. M. Monaghan & L. C. Smith (2004). Dairy pasture responses to soil physical properties. *Australian Journal of Soil Research*, 42, 99-105.

Drewry, J. J., R. J. Paton & R. M. Monaghan (2004). Soil compaction and recovery cycle on a Southland dairy farm: implications for soil monitoring. *Australian Journal of Soil Research*, 42, 851-856.

Edwards, W. M., M. J. Shipitalo, S. J. Traina, C. A. Edwards & L. B. Owen. (1992). Role of Lumbricus terrestris (L.) burrows on the quality of infiltrating water. *Soil Biology & Biochemistry*, 21, 1555-1561.

Ghani, A., B. M. Dexter & K.W. Perrott (2003). Hot-water extractable carbon in soils: a sensitive measurement for determining impacts of fertilisation, grazing and cultivation. *Soil Biology & Biochemistry*, 35, 1231-1243.

Goh, K. M. & P. H. Williams (1999). Comparative nutrient budgets of temperate grazed pastoral systems. *In:* E. M. A. Smaling, O. Oenema and L. O Fresco (eds), *Nutrient Disequilibria in Agroecosystems: Concepts and Case Studies.* CABI Publishing, Wallingford, 265-294.

Gregorich, E. G., M. R. Carter, D. A. Angers, C. M. Monreal & B .H. Ellert (1994). Towards a minimum data set to assess soil organic matter quality in agricultural soils. *Canadian Journal of Soil Science*, 74, 367-385.

Haygarth, P. M., P. J. Chapman, S. C. Jarvis & R. V. Smith (1998). Phosphorus budgets for two contrasting farming systems in the UK. *Soil Use and Management*, 14, 160-167.

Haygarth, P. M. & S. C. Jarvis (1999). Transfer of phosphorus from agricultural soils. *Advances in Agronomy*, 66, 195-249.

Haynes, R. J. & P. H. Williams (1992). Accumulation of soil organic matter and the forms, mineralization potential and plant-availability of accumulated organic sulphur: Effects of pasture improvement and intensive cultivation. *Soil Biology & Biochemistry* 24, 209-217.

Haynes, R. J. & P. H. Williams (1993). Nutrient cycling and soil fertility in the grazed pasture ecosystem. *Advances in Agronomy*, 49, 119-199.

Hooda, P. S., D. McNulty, B. J. Alloway & M. N. Aitken (1997). Plant availability of heavy metals previously amended with heavy applications of sewage sludge. *Journal of Science, Food and Agriculture*, 73, 446-454.

Jarvis, S. C., D. Scholefield & B. Pain (1995). Nitrogen cycling in grazing systems. *In:* PE Bacon (ed), Nitrogen fertilisation in the environment. Marcel Dekker, New York, 381-419.

Lambert, M. G., D. A. Clark, A. D. Mackay & D. A. Costall (2000). Effect of fertiliser application on nutrient status and organic matter content of hill soils. *New Zealand Journal of Agricultural Research,* 43, 127-138.

Lambert, M. G., B. P. Devatier, P. Nes & P. E. Penny. (1985). Loses of nitrogen, phosphorus, and sediment in runoff from hill country under different fertiliser and grazing management regimes. *New Zealand Journal of Agricultural Research,* 28, 371-379.

Larson, W. E. & F. J. Pierce (1994) The Dynamics of Soil Quality as a Measure of Sustainable Management. *In:* J. W. Doran, D. C. Coleman, D. F. Bezedick & B. A. Stewart, (eds), *Defining Soil Quality for a Sustainable Environment,* Special Pub. No. 35; American Society Agronomy: Madison, Wisconsin, 37-51.

Ledgard, S. F. (2001). Nitrogen cycling in low input legume-based agriculture, with emphasis on legume/grass pastures. *Plant and Soil,* 228, 43-59.

Ledgard, S. F., S. C. Jarvis & D. J. Hatch (1998). Short-term nitrogen fluxes in grassland soils under different long-term nitrogen management regimes. *Soil Biology & Biochemistry,* 30, 1233-1241.

Maquire, R. O. & J. T. Sims (2002). Soil testing to predict phosphorus leaching. *Journal of Environmental Quality,* 31, 1601-1609.

McDowell, R. W. & A. N. Sharpley (2001). Approximating phosphorus release from soils to surface runoff and subsurface drainage. *Journal of Environmental Quality,* 30, 508-520.

McDowell, R. W., A. N. Sharpley, P. C. Brookes & P. R. Poulton (2001). Relationships between soil test phosphorus and phosphorus release to solution. *Soil Science,* 166, 137-149.

McGarry, D. & G. Sharp (2001). A rapid, immediate, farmer-usable method of assessing soil structure condition to support conservation agriculture. *In:* Proceedings of the 1st World Congress on Conservation Agriculture, Madrid, Spain, Vol. 2, 209-214.

Metherel, A. K. (2003). Management effects on soil carbon storage in New Zealand pastures. *In:* Proceedings of the New Zealand Grassland Association, 65, 259-264.

Naeth, M. A., D. J. Pluth, D. S. Chanasyk, A. W. Bailey & A. W. Fedkenheuer (1990). Soil compaction impacts of grazing in mixed prairie and fescue grassland ecosystems of Alberta. *Canadian Journal of Soil Science,* 70, 157-167.

Oenema, O., G. L. Velthof, S. Yamulki & S. C. Jarvis (1997). Nitrous oxide emissions from grazed grassland. *Soil Use and Management,* 13, 288-295.

Okano, S., K. Sato & K. Inoue (1991). Turnover rate of soil biomass nitrogen in root mats layer of pastures. *Soil Science and Plant Nutrition,* 33, 373-386.

Olsen, S. R. & L. E. Sommers (1982). Phosphorus. *In:* A. L. Page *et al.* (eds), Methods of Soil Analysis, part 2. *Agronomy,* 9, 402-430.

Parfitt, R. L., G. W. Yeates, D. J. Ross, A. D. Mackay & P. J. Budding (2005). Relationships between soil biota, nitrogen and phosphorus availability, and pasture growth under organic and conventional management. *Applied Soil Ecology,* 28, 1-13.

Parr, J. F. (1974). Pesticides and microorganisms in soil and water. *In:* W.D. Guenzi (ed), Pesticides in Soil and Water. Soil Science Society of America, Madison, WI., 315-316.

Peoples, M. B. & J. A. Baldock (2001). Nitrogen dynamics of pastures: Nitrogen fixation, the impact of legumes on soil nitrogen fertility, and the contribution of fixed N to Australian farming systems. *Australian Journal of Experimental Agriculture,* 41, 327-346.

Peoples, M. B., J. F. Angus, A. D. Swan, B. S. Dear, H. Hauggaard-Nielsen, E. S. Jensen & M. H. Ryan (2004). Nitrogen dynamics in legume-based pasture systems. *In:* A. R. Mosier, J. K. Syers & J.R. Freney (eds), *Agriculture and the Nitrogen Cycle.* Island Press, Washington,103-114.

Rasse, D. P., C. Rumpel & M-F. Dignac (2004). Is soil carbon mostly root carbon? Mechanism for a specific stabilisation. *Plant and Soil,* 232, 1-16.

Saggar, S. (2004). Changes in nitrogen dynamics of legume-based pastures with increased nitrogen fertiliser use: Impacts on New Zealand's nitrous oxide emissions inventory. *New Zealand Soil News,* 52, 110-115.

Sarathchandra, S. U., A. Lee, K. W. Perrott, S. S. Rajan, E. H. A. Oliver & I. M. Gravett (1994). Effects of phosphate fertiliser applications on microorganisms in pastoral soils. *Australian Journal of Soil Research,* 31, 299-309.

Scott, A., B. C. Ball, I. J. Crichton & M. N. Aitken (2000). Nitrous oxide and carbon dioxide emissions from grassland amended with sewage sludge. *Soil Use and Management,* 16, 36-41.

Sharpley, A. N. & J. K. Syers (1976). Potential role of earthworm casts for the phosphorus enrichment of run-off waters. *Soil Biology & Biochemistry,* 8, 341-346.

Shepherd, T. G. (2000). Visual soil assessment: Volume 1. Field guide for cropping and pastoral grazing on flat to rolling country. Horizons.mw/Landcare Research, Palmerston North, 84 p.

Sims, J. T., A. C. Edwards, O. F. Schoumans & R. R. Simard. (2000). Integrating soil phosphorus testing into environmentally based agricultural management practices. *Journal of Environmental Quality,* 29, 60-71.

Smith, P., D. S. Powlson, P. U. Smith, P. Fallon & K. Coleman (2000). Meeting the U.K.'s climate change commitments: options for carbon mitigation on agricultural land. *Soil Use and Management,* 16, 1-11.

Sparling, G. & L. Schipper (2004). Soil quality monitoring in New Zealand: trends and issues arising from a broad-scale survey. *Agriculture, Ecosystem & Environment,* 104, 545-552.

Springett, J. A. (1983). Effect of five species of earthworm on soil properties. *Journal of Applied Ecology*, 20, 865-872.

Steen, E. (1991). Usefulness of the mesh bag method in quantitative roots studies. *In:* D. Atkinson (ed). *Ecological Perspectives*, Blackwell Scientific Publications, London, 75-86.

Stewart, D. P. C. & A. K. Metherell. (1999b). Carbon (^{13}C) uptake and allocation in pasture plants following field pulse labelling. *Plant and Soil*, 210, 61-75.

Stockdill, S. M. J. (1982). Effects of introduced earthworms on the productivity of New Zealand pastures. *Pedobiologia*, 24, 29-35.

Tate, K. R., D. J. Ross, A .J. Ramsay & K. N. White (1991). Microbial biomass and bacteria in two pasture soils: an assessment of measurement procedures, temporal variation and the influence of P fertility status. *Plant and Soil*, 132, 233-241.

Tate, K. R., D. J. Giltrap, J. J. Claydon, P. F. Newsome, I. A. Atkinson, M. D. Taylor & R. Lee (1997) Organic carbon stocks in New Zealand's terrestrial ecosystems. *Journal of Royal Society of New Zealand,* 27, 315-335.

Thomas S. M., G. S. Francis, H. E. Barlow, M. H. Beare, L. A. Trimmer, R. N. Gillespie & F.J. Tabley (2004). Winter grazing of forages - soil moisture and tillage effects impact nitrous oxide emissions and dry matter production. *Proceedings of the New Zealand Grassland Association,* 66, 135-140.

Towers, W. & P. Horne (1997). Sewage sludge recycling to agricultural land: the environmental scientist's perspective. *Journal of the Institute of Water and Environmental Management*, 11, 126-132.

Wheeler, D. M., G. P. Sparling & A. H. C. Roberts (2003). Using soil test databases to monitor benchmark changes in soil fertility over time. *In*: L. D. Currie & J. A. Hanly (eds), *Tools for nutrient and pollutant management: Applications to agriculture and environmental quality*, Occasional Report No. 17. Fertiliser and Lime Research Centre, Massey University, Palmerston North, New Zealand, 349-359.

Willems, J. J. G. M., J. C. Y. Marinissen & J. M. Blair (1996). Effects of earthworms on nitrogen mineralization. *Biology & Fertility of Soils*, 23, 57-63.

Physical constraints in grassland ecosystems

I.M. Young[1], K. Ritz[2], C.S. Sturrock[1] and R. Heck[3]

[1]SIMBIOS Centre, University of Abertay, Bell Street, Dundee, Scotland, DD1 1HG, UK, Email: imy@tay.ac.uk, [2]National Soils Resource Institute, Cranfield University, Silsoe, MK45 4DT, UK, [3] Department of Land Resource Science, Ontario Agricultural College, University of Guelph, Guelph, Ontario, Canada N1G 2W1

Key points

1. Grassland system management must account adequately for biophysical and biochemical processes.
2. Amenity turf, especially sports turf, provides an excellent case study on how biology and physics interact to impact on the functioning of a grass ecosystem.
3. Microorganisms produce hydrophobic compounds that act to decrease water ingress and alter the function of the soil-grass ecosystem.

Introduction

One of the greatest challenges in soil ecology is to integrate the myriad of component parts of the ecosystem to provide a better understanding of the whole system in terms of appropriate management procedures. Great strides have been made in many areas of production-orientated intensive agriculture that rely on the ready availability of cheap and efficient means to control the nutritional requirements of the crop and its biological environment. This is as true for grassland soils as it is for soils under arable crops. However, three issues now bear down heavily on soil ecosystem management, as the focus shifts from production-based agriculture to one set firmly in the context of wider environmental management. Notwithstanding ecological common sense, policy and legislative drivers for this are rapidly emerging in Europe and elsewhere. Firstly, we have to quantify how a certain use and management regime of an ecosystem impacts on the wider environment. This will involve a better understanding of how to predict energy flows across a range of ecosystems, and thus scales. Secondly, the intensive management systems that are currently widespread in temperate agriculture are under pressure to change to a less intensive approach with appropriate protections for the wider environment. Thus, the use of many synthetic chemicals is now being deemed inappropriate and illegal in certain important cases. We must now address a better understanding of the natural soil ecosystems and processes therein. Finally, we are slowly coming to realise that any such understanding of soil ecosystems demands an integrative approach that bridges across disciplines rather than the solution residing solely in a chemistry, biology or physics lab, or indeed in a computer processor.

With this in mind we have doubts over the usefulness of looking solely at the physical constraints of a system without taking into account the integrative processes that are involved across the soil ecosystem. In this paper we will draw on a case study that emphasises these processes.

The basic generic characteristics of the physical framework of the soil supporting a grass based ecosystem are that it must be loose enough to permit adequate root growth, firm enough to allow vehicular and animal traffic, and allow water to drain at a sufficient rate that minimises water logging, but permit adequate retention of water and solutes for plant development and soil-based biologically mediated processes. It must also be stable enough to resist deformation or erosion that would significantly impact on the main functioning of the ecosystem. These, somewhat contradictory, characteristics actually hold for any plant system, growing in any soil.

A world perspective

Natural grasslands can be found on almost all continents, comprised primarily by the Great Plains of North America, Pampas of South America, Puszta of Eastern Europe, Steppe of Russia, Veld of Africa, as well as the grasslands of the Sahel, Mongolia, central Asia and of Australia. The climate of these regions is characterised by a semi-arid climate, where the rates of precipitation and potential evapotranspiration are approximately equal. As the climate becomes more arid, potential evapotranspiration exceeding precipitation, the vegetation changes to shrubland and desert. In more humid climates, the vegetation changes to savannah and forests.

The dominant process of soil development is the development of dark, organic matter-rich surface A horizon. This is favoured by the regular incorporation (predominantly sloughing of fibrous roots) of fresh organic residues, in conjunction with soil conditions favouring its humification and retention. Of particular importance is the role of calcium ions in binding organic matter to the surfaces of soil minerals, reducing its vulnerability to microbial decomposition (for this reason the term 'calcification' has been applied to this process). Anderson et al. (1981) found that about half of the organic matter in grassland soils was associated with the clay fraction. Over time, the development of the fibrous root system and incorporation of organic residues leads to stable aggregation of primary particles, reducing the potential for translocation of finer particles in the soil profile. Typically, the dominant control on vegetative production in grasslands is moisture, therefore the development of the A horizon, increases from arid through semi-arid to humid climate (until the dominant vegetation becomes forest). Under intense tillage, a considerable portion of this organic matter can be lost (Rennie and Clayton, 1967); this loss results in aggregate destabilisation, with consequent surface compaction as well as mobilisation of fine particles and subsurface densification.

Moderate to well drained soils in the semi-arid or more humid climates, are typically leached of soluble chloride and sulphate salts. Where drainage is impeded, or the climate is more arid, soluble salts can accumulate at the soil surface by capillary movement of water in response to removal by evapotranspiration. The source of these salts maybe the soil parent material (marine deposits), deposition of ocean spray, in situ weathering of minerals, or redistribution through regional water cycles. If the accumulated salts are high in sodium, the proportion of sodium ions occupying the cation exchange complex of minerals and organic colloids increase (solonisation). Once exchangeable sodium reaches about 15%, the clay sized particles become susceptible to dispersion and hydrolysis (reflected by pH values above about 8.5). When precipitation (rain) occurs, dispersed particles can be easily translocated to underlying layers (B horizons), progressively blocking smaller pores and coating the surfaces of aggregates. Such natric B horizons are characterised by impeded soil hydraulic conductivity and aeration. The elevated pH, also accelerates the decomposition of primarily silicate minerals; secondary siliceous minerals are usually deposited near the top of the natric B horizon. Strongly developed natric B horizons exhibit columnar aggregation, that is very hard when dry. Root development, and infiltration of moisture become concentrated in the inter-aggregate voids, making the vegetation more susceptible to drought. Over time, the surface horizon becomes depleted of both colloidal material and Na, resulting in an acid eluvial horizon (solodisation). Aggregation in the eluviated horizon is often platy in form and subject to slaking.

Ultimately, moderate to well drained soils are also subject to leaching of minerals, which are more stable than the chlorides and sulphates, either in dissolved, or suspended form. The translocation of primary carbonates typically occurs by dissolution; if inadequate moisture is available, secondary carbonates may precipitate at the base of the soil profile. Carbonate accumulation in soils developed from non-calcareous parent materials can be related to the weathering of silicate minerals in the presence of carbonated soil solution (dissolved carbon dioxide from the atmosphere or metabolic respiration in the soil). Continued accumulation of secondary carbonates contributes to blockage of porosity and, in more extreme circumstances may result in cementation of the illuvial layer.

Unaggregated particles, especially clay sized, are also subject to translocation (a process called lessivage). In coarse-textured parent materials, suspended particles may be translocated to considerable depths, possibly not being deposited until the watertable is reached; in such circumstances, numerous thin illuvial lenses may develop, until the soil profile is devoid of the finer particles. In soils developed from parent materials richer in fine particles, the depth of deposition becomes shallower and the thickness of the illuvial layer is often greater. In fine-textured parent materials, translocated particles would typically be 'filtered' very quickly and the illuvial layer develops at very shallow depths. When sufficient clay is present in the parent material, desiccation leads to vertical cracking and the formation of prismatic aggregates; over time the surfaces of such aggregates become coated with eluviated colloids. It is important to note that clay-rich layers or horizons can also be inherited from the parent material (varied sedimentation) or weathering of primary minerals and neo-formation of secondary layer silicates.

The filling of subsurface voids, by translocated particulates or precipitation of secondary soil minerals, can significantly alter the hydraulic characteristics of a soil. In extreme cases, the rate of water infiltration may be so retarded that a perched watertable may form. As the diffusion of gases is very slow in saturated soil, the horizons above the illuvial impediment, quickly become anaerobic (gleyed) in response to continued microbial metabolism. Such conditions favour the chemical reduction and mobilisation of iron, which may coalesce and, depending on the supply of iron in the parent material, lead to localised or more extensive cementation through the precipitation of secondary iron oxides.

Just as regional trends in moisture efficiency are manifest in the strength of specific soil processes, localised redistribution of water can also impart a very profound influence. In undulating landscapes, the knolls or ridges are typically more arid than the side slopes, because of both shedding of precipitation, greater height above any groundwater table and greater exposure to winds. At the other extreme, depressions and swales tend to be more humid, also because accumulation of precipitation, proximity to groundwater table as well as sheltering from wind. In high latitudes, equatorial and west facing slopes also experience greater heating, than polar and east facing slopes, and so are more arid. Consequently, along a single cantena, it is possible to encounter large gradients in soil development, from the perspectives of organic matter accumulation and A horizon development, leaching or accumulation of soluble salts (and the possibility of solonization-solodization), the leaching and possible lateral redistribution of carbonates, the formation of clay-rich horizons (by redistribution or neo-formation), as well as gleying.

In areas where the parent material contains large amounts of swelling clay minerals (smectites), the cyclic wetting and drying associated with grasslands, can result in considerable contraction of the soil mass, opening cracks up to centimetres of width and meters of depth. Surface material may be introduced to these cracks by gravity, wind, water or faunal activity. When rewetted the soil dilates, the extra material at the bottom of the cracks, creates stress which is relieved by upward swelling of the soil; recognised by mounding (called gilgai) on the soil surface. Over repeated cycles there is a slow, but definite churning of the soil mass corresponding with depth of cracking. Such churning inhibits the normal development of horizons. Soils in which this process has dominated their development are most commonly called Vertisols. Less accentuated vertic behaviour can also be recognised in soils with smaller amounts of swelling clays, or less intense wetting and drying cycles.

In natural and semi-natural grassland systems the opportunity to perturb the system to relieve any physical constraints is limited. The main physical constraints in grassland systems relate to water (too much or too little), and carbon inputs. The former are ameliorated by irrigation or drainage, where possible, and the latter, linked directly to the stability of soil, is perceived to be of growing importance in terms of carbon sequestration. According to Conant *et al.* (2001), grassland contains 12% of the earth's soil organic matter, which averages 331 Mg/ha in temperate systems. Driven primarily by funding priorities, much work is progressing in this area. Although here we add a note of caution: unless there are long-term plans to encourage land users to maintain areas of land under grass then any effect of sequestration will rapidly be lost if subsequent alterations in land-management practices do not result in their persistence. One without the other is of little use in the long term, and there is no consistent analysis of the upper limit of carbon that we can pragmatically expect to capture in grasslands.

In terms of soil stability, grassland systems have consistently exhibited the highest resilience against erosive forces, and there is little reason to persist in examining such constraints, except perhaps to further investigate fundamental cohesion-adhesion properties between biota and mineral components of the soil. Carbon inputs again are seen as a key mechanism here (Balabane & Plante, 2004).

Opportunities to ameliorate physical constraints in grasslands are far more prevalent in intensively used grasslands such as Amenity grass and Sports Turf. In many cases intensive management occurs on a daily basis, and it is to this type of grassland that we turn for the remaining sector of the paper.

Golf greens: a case study in intensively managed grass ecosystems

The soil-grass bionetwork found in golf greens provides an example of the extreme end of intensively managed grasslands, and hence a useful case study on how knowledge across a range of scientific disciplines can interact to produce a functional grassland with such precise specifications.

The most recognised grassland system in the world is the Old Course at St Andrews in Scotland. Within a one-hour drive of this course there are over 100 other golf courses, with 20 courses within a 10-minute drive. In Scotland, there are 537 courses covering 23,000 hectares that contain 33 designated Sites of Special Scientific Interest (SSSI). Around the World there are over 8000 courses, with hundreds more being constructed annually, especially in China and Africa. Whilst in terms of total grassland, golf courses make but a small percentage of the total area, economically they are seen to be more important than natural or semi-natural grasslands, and importantly they have greater resonance (both positive and negative) with the public and politicians.

Taking new green constructions as an example. The physical habitat of such greens has a strict prescription based on the sand grain size, root penetration, and the saturated hydraulic conductivity, which aims to produce a homogenised (and thus easily manageable) green. At the start of a green's life it may be made up of > 98% sand and <2% organic matter: this ratio can vary markedly but as we shall see sand plays an important role in any course. Such composition is based on the ecosystem's ability to deal with water and solutes. N applications may range from none, to very low (30 kg/ha) in the traditional Links courses, to very high (260 kg/ha) in new constructions. The application of N to most new courses makes up for the low N content and cation exchange capacity of the sand based ecosystem. This initially maximises grass coverage and the look of the green surface, which is an important commodity in golf, in the same way that yield is an important commodity in agricultural ecosystems.

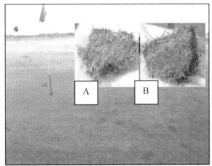

Figure 1 Dark patches on golf green reveal patchy fungal damage to turf. Inset image (A) provides a close-up view of turf and near surface soil. This shows the soil-turf complex with fungal damage, compared to inset image (B) which was sampled within 5 cm of (A). Typically such damage is also related to increases in water repellency, and affects the putting surface of the golf green and water ingress. IMAGE: Iain M Young, SIMBIOS Centre

In addition, every year most Course Managers dress each green with more sand. Thus, even in traditional greens that started life with, for example, a loam texture, over perhaps a few short years this will have dramatically altered simply due to the addition of sand, that is a characteristic of most course management practices, or the build-up of thatch from dead roots at the surface. In practice, what appears to happen is that despite the ecosystem being designed to minimise physical constraints, a whole new set of biophysical challenges arise. The prescribed structures act as almost ideal incubators for fungal and bacteria populations (pathogenic and non-pathogenic). Additionally, the system, built to maximise homogeneity, soon reveals itself to have a significant heterogeneous character. Where fungal pathogens, or nutritional stresses, invade the green they do so in a patchy fashion (Figure 1). What this reveals is the importance of small-scale variations in the characteristics of ecosystems that, at the scale we manage most things at, are incredibly difficult to predict, even within what is apparently a 'uniform' system. Even slight variations in say the mixing procedures of sand and organic matter during construction, or variations in sand composition and shape, will produce a significant spatially and temporally heterogeneous habitat. These small-scale variations will be reinforced over time spreading to larger scales and/or immediately force biota to operate in a similar variable fashion. Thus, at larger spatial scales, over reasonably short time scales (months) heterogeneity is observed, simply because of the presence of small-scale heterogeneity at the birth of the ecosystem.

The interesting point here is that the physics of the ecosystem rather than acting as a constraint, acts as a framework through which all biota must operate and all energy flows (i.e. water, gas, dry matter) must pass. Research on such high-end amenity ecosystems has revealed another interesting interaction between biological and physical interactions. Taking again the example of golf greens, but the case holds for many soil-plant systems, it has been observed that the surface of the greens may become resistant to water ingress, despite the dominant sand composition. In many courses adding surfactants to decrease the surface tension and contact angle of water, and thus increase the rate of water ingress, has become a common management practice. The question is why, in a sand dominated ecosystem, do hydrophobic conditions appear?

Two possible, perhaps interacting, processes are involved. Firstly, the spread of fungi through soil systems is partially controlled by the air-filled porosity and the pore connectivity (Harris et al., 2001). Unlike bacteria, soil fungi do not require continual water-films to remain active (Ritz & Young, 2004). As fungal growth is focused within such air-filled pores, and can spread across the surfaces of soils, they are exposed to relatively severe drying conditions and thus they have developed a number of strategies to minimise desiccation. One involves the deposition of hydrophobic substances such as melanin (Nosanchuk & Casadevall, 2003) and a variety of hydrophobins (Wosten, 2001) in the outer walls, which are inevitably also deposited into the soil. Within the confines of tight pore spaces in soil, such hydrophobic chemicals will be spread at the edges of the pore surface onto the mineral soil rendering them increasingly repellent to water ingress. Recently, Wright & Upadhyaya (1996) have postulated that AM fungi release copious quantities of a hydrophobic glycoprotein called glomalin that acts to slow the rate of water ingress and thus engenders habitat stability by reducing a process known as slaking (Griffiths & Young, 1994). As AM fungi are an important part of the microbial community in any grassland it may be that glomalin is an important integrator within the grassland ecosystem. However, much research is still required to provide a full account of the role of glomalin in habitat stability (Feeney et al., 2004). Whatever the evidence reveals with respect to glomalin, there are observed cases of fungi producing a range of hydrophobic compounds in soil systems that impact on the water flow. Related to this, we have clear evidence of the role of soil microbes in improving structural stability through the exudation of extracellular polysaccharides (Preston et al., 2001). Secondly, much of the root growth in grasses in most soil ecosystems occurs within the first 5 cm of the soil. In golf greens this can lead to a build up of thatch that again acts independently as a barrier to water ingress (Taylor & Blake, 1982), and as a highly suitable environment for fungi to explore to access substrate.

Thus, the interactions between biological and physical components play a key role in this ecosystem, which drive the creation of a heterogeneous habitat over relatively short time-scale and impact directly on the functionality and the management of the system.

Conclusions

The short case study presented has many parallels with natural and managed grasslands. The simplicity and importance of the golf green is that you can literally monitor the birth of the soil-grass ecosystem and chart progress over time, quantifying how the biophysical network impacts on function. We concentrate on hydrophobicity that is present in more natural systems (Naeth *et al.*, 1991), and the issues such as compaction, drainage, and pest control are prevalent, and important, in both systems.

References

Anderson, D. W., S. Sagger, J. R. Bettany & J. W. B. Stewart (1981). Particle size fractions and their use in studies of soil organic matter: I. The nature and distribution of forms of carbon, nitrogen and sulfur. *Soil Science Society America Journal,* 45, 767-772.

Balabane M. & A. F. Plante (2005). Aggregation and carbon storage in silty soil using physical fractionation techniques. *European Journal of Soil Science*, 55, 415-427.

Conant R. T., K. Paustian & E. T. Elliot (2001). Grassland management and conversion into grassland: effects on soil carbon. *Ecological Applications*, 11, 343-355.

Feeney, D. S., T. Daniell, P. D. Hallett, J. Illian, K. Ritz & I. M. Young (2004). Does the presence of glomalin relate to reduced water infiltration through hydrophobicity? *Canadian Journal of Soil Science*, 84, 365-372.

Griffiths, B. S. & I. M. Young, I. M. (1994). The effects of soil-structure on protozoa in a clay-loam soil. *European Journal of Soil Science*, 45, 285-292.

Harris, K., I. M. Young, C. A. Gilligan, W. Otten & K. Ritz (2003). Effect of bulk density on the spatial organisation of the fungus *Rhizoctonia solani* in soil. *FEMS Microbiology Ecology*, 44, 45-65.

Naeth, M. A., A. W. Bailey, D. S. Chanasyk & D. J. Pluth (1991). Water holding capacity of litter and soil organic matter in mixed prairie and fescue grassland ecosystems of Alberta. *Journal of Range Management*, 44, 13-17.

Nosanchuk J. D. & A. Casadevall (2003). The contribution of melanin to microbial pathogenesis. *Cellular Microbiology*, 5, 203-223.

Preston, S., S. Writh, K. Ritz, B. S. Griffiths .S. & I. M. Young, I.M. (2001). The role played by microorganisms in the biogenesis of soil cracks: importance of substrate quantity and quality. *Soil Biology & Biochemistry*, 33, 1851-1858.

Rennie, D. A. & J. S. Clayton (1967). An evaluation of techniques used to characterize the comparative productivity of soil properties in Saskatchewan. *Trans. Comm. II and IV, ISSS,* Aberdeen, pp. 365-376

Ritz, K. & I. M. Young (2004). Interactions between soil structure and fungi. *Mycologist*, 18, 51-59.

Taylor, D. H. & G. R. Blake. (1982). The effect of turfgrass thatch on water infiltration rates. *Soil Science Society America Proceedings*, 46, 616–619.

Wosten, H.A.B. (2001). Hydrophobons: Multipurpose proteins. *Annual Review of Microbiology*, 55, 625-646.

Wright, S. F. & A. Upadhyaya (1996). Extraction of an abundant and unusual protein from soil and comparison with hyphal protein of arbuscular mycorrhizal fungi. *Soil Science*, 161, 575-586.

Integrating below-ground ecology into sustainable grassland management

R.D. Bardgett

Institute of Environmental and Natural Sciences, Lancaster University, Lancaster LA1 4YQ, UK, Email: r.bardgett@lancaster.ac.uk

Key points

1. Grasslands produce soils that sustain an abundant and diverse soil food web, providing tremendous opportunity for below-ground interactions to influence nutrient cycling processes and plant production.
2. Fast developing areas of ecological science offer scope to harness positive outcomes of below-ground ecology for enhancing efficient cycling of nutrients in sustainable grassland systems.

Keywords: nitrogen cycling, soil biota, organic nitrogen, mycorrhizal fungi, soil microbes, grazing animals, root exudation

Introduction

Grasslands, including steppes, savannas, and prairies, are important terrestrial ecosystems covering about a quarter of Earth's land surface. The development of agriculture has been very closely linked to these grasslands, and they now form the backbone of the global ruminant livestock industry, completely dominating the landscape of many parts of the world. In recent years, the nature of grassland agriculture in many parts of the world has started to change. Until recently, the drive was to maximise production through the use of large amounts of fertiliser and nitrogen (N) responsive grasses. Now, the need to develop sustainable management strategies that encourage efficient nutrient cycling, thereby minimising the use of fertilisers, has come to the fore. This has largely been driven by agricultural policy aimed at cutting fertiliser use to reduce nutrient losses to air and water, and also to help restore botanical diversity to species-poor agriculturally improved grassland. There is also an increasing drive towards organic production systems, which rely entirely on biological processes of nutrient cycling to meet crop demand. Collectively, these demands highlight the urgent need to develop management strategies for grassland that are directed at encouraging a greater reliance on natural ecological processes, rather than on the input of artificial inputs such as fertilisers.

Grassland lends itself well to management for natural processes of nutrient cycling. This is because they build soil systems that are very different from those of other vegetation types; their high turnover of shoot and root biomass creates a large pool of labile organic matter at the soil surface, and heavy herbivore loads result in large amounts of organic matter being returned to soil as animal waste, which is nutrient replete. These features combine to produce a soil environment that sustains an abundant and diverse soil food web, the main agent of efficient nutrient cycling; hence, there is a tremendous opportunity for soil organisms and their interactions with plants, and their consumers, to influence nutrient cycling processes and plant production. This is what this paper is concerned with: illustrating, using selected examples, how an understanding of ecological interactions between plants and soil biota can be harnessed to promote efficient nutrient cycling and sustainable management in agricultural grassland. I will use three examples from rapidly developing areas of research to demonstrate this. First, I will consider recent developments in the area of plant nutrition that reveal the potential for grassland plants to use organic N forms, as opposed to a complete reliance on mineral N. Second, I will discuss recent studies that show intimate links between grazers, plants, and soil microbes, which have important consequences for plant production in grazed grassland. Finally, I will consider research that reveals how management of grassland can be altered to maximise soil biodiversity, especially the fungal component of soils, with associated benefits for plant communities. The examples that I use have not yet been integrated into grassland management; the aim here is to illustrate how knowledge of these ecological interactions might be harnessed in sustainable management.

Case 1: A role for organic nitrogen in grassland systems

Historically, a central assumption of terrestrial nutrient cycles was that for soil N to be available for plant uptake it needs to be in an inorganic form. A growing body of evidence now challenges this view, pointing to the importance of dissolved organic nitrogen (DON) in the form of amino acids for plant nutrition. This is especially the case in strongly N limited ecosystems, such as alpine and arctic tundra, where studies reveal that plants can take up amino acids directly, by-passing the need for microbial mineralisation to produce simpler inorganic N forms (Chapin et al., 1993; Kielland, 1994; Schimel & Chapin, 1996; Lipson & Monson, 1998; Raab et al., 1999; Henry & Jefferies, 2003a, b). This is especially significant since DON often represents the dominant form of soluble N in these ecosystems (Kielland, 1994; Jones & Kielland, 2002), thereby providing a previously unrecognised source of N for plant growth.

Whilst most research in this area has focussed on natural ecosystems, recent work reveals that DON may also be of significance for plant nutrition in agricultural settings. There are two lines of evidence to support this. First, studies of temperate grassland reveal that amino acids can reach significant concentrations in soil, sometimes reaching equal abundance to mineral N (Bardgett et al., 2003). Second, studies done under both laboratory and field conditions show that grassland plants have the potential to take-up amino acids from soil directly, as in natural settings (Näsholm et al., 2000; Streeter et al., 2000; Bardgett et al., 2003; Weigelt et al., 2005). This is not a straightforward issue, however. For example, it has also been argued that due to fast microbial turnover of organic N in agricultural soils, the main route for plant uptake of N is as mineral N after microbial mineralisation (Hodge et al., 1998, 1999; Owen & Jones, 2001). That organic N uptake is of limited importance in agricultural situations is also evidenced by in situ measurements of N uptake by plants in temperate grasslands, which show that whilst glycine can be taken up directly by plants, microbial turnover and release of this N into the plant-soil system is the major pathway for N acquisition (Bardgett et al., 2003). Despite this, field studies do reveal that amino acids could be of some significance for plant nutrition in low productivity, unfertilised grasslands where amino acids are relatively abundant in soil (Bardgett et al., 2003). This implies, therefore, that there is potential for organic N use in certain situations; hence, the challenge is to identify these situations and management strategies that best optimise organic N availability in soil.

Understanding organic N cycling may also be of significance in other areas of grassland agriculture, especially in relation to promoting plant diversity. Whilst the capacity of plants to take up organic N directly appears to be ubiquitous, there is emerging evidence that plants vary in their capacity to take up different chemical forms of N, showing species-specific preferential uptake of either organic or inorganic N. So far, species-specific differences have been demonstrated most clearly in arctic tundra (McKane et al., 2002) and in alpine communities (Miller & Bowman 2002, 2003). Studies of grassland plants also indicate a degree of preferential uptake of different forms of N by some species, albeit under laboratory conditions (Weigelt *et al.*, 2005). These findings are potentially significant since the existence of species-level differences in the preference of plants to take up chemical forms of N could provide a mechanism for plants to efficiently partition a limited soil N pool, thereby facilitating species coexistence and the maintenance of plant diversity (McKane *et al.*, 2002) (Figure 1). The implication here, therefore, is that managing to promote diversity of N forms in soil could contribute to the maintenance of plant diversity, a key objective of sustainable grassland management. A similar model could also be envisaged for P, which similarly occurs in many forms in soil, both organic and inorganic (Turner *et al.*, 2004). It is important to stress, however, that evidence for this is still lacking and work is urgently needed to test the importance of resource partitioning for plant communities in both natural and agricultural settings.

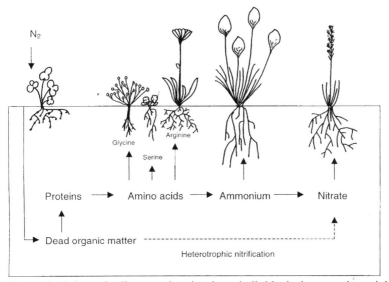

Figure 1 Schematic diagram showing how individual plant species might access different chemical forms of N enabling species to coexist in terrestrial ecosystems (redrawn from Bardgett, 2005)

The above studies all suggest that direct uptake of DON by plants could be of importance for nutrient cycling and vegetation structure in grassland, but especially in unfertilised grassland systems, which are the cornerstone of sustainable agriculture. However, there is still much more work to be done in this area. This point was recently emphasised by Jones *et al.* (2005) who argued that experimental approaches used to assess whether DON is important may be compromised because of the use of inappropriate methods for comparing and quantifying plant available inorganic and organic soil N pools. In addition, they argued that experiments aimed at quantifying plant DON capture, which typically use dual-labelled (N-15, C-13) organic N tracers, often do not consider important aspects such as isotope pool dilution, differences in organic and inorganic N pool turnover times, bi-directional DON flows at the soil-root interface, and the differential fate of the ^{15}N and ^{13}C in the tracer compounds. Based upon experimental evidence, they hypothesised that DON uptake from the soil may not contribute largely to N acquisition by plants but may instead be primarily involved in the recapture of DON previously lost during root exudation. In summary, the jury is still out and further work is needed to establish the importance of amino acids as plant N acquisition pathway in grassland systems.

Case 2: Links between grazers, plants and soil microbes

Grazing animals are integral to grassland agriculture, and it has long been recognised that they can have important effects on both the productivity and vegetation composition of grassland (Bardgett & Wardle, 2003). However, only recently have ecologists started to consider how grazers can influence decomposer communities, and the consequences of this for grassland production and livestock carrying capacity. A central tenant of ecology is that grazers can actually increase plant production when at intermediate densities (McNaughton, 1985). This is especially the case in areas of high soil fertility, such as productive grassland ecosystems, where the phenomena is refereed to as *compensatory growth* (De Mazancourt *et al.,* 1999). Various mechanisms have been proposed to explain this, especially the priming of soil nutrient cycling through the recycling of plant material as animal waste - which is nutrient replete - rather than through a plant litter pathway (Bardgett *et al.*, 1998). Recent studies, however, point to other mechanisms that involve soil microbes and roots that could be of high significance for sustainable management.

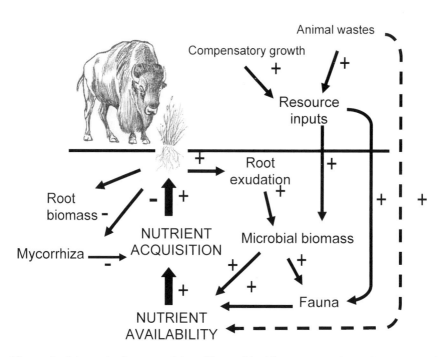

Figure 2 Schematic diagram of the effects of herbivory on producer-decomposer feedbacks in fertile grasslands that result from changes in the quantity of resources returned to the soil. These mechanisms are commonest in nutrient rich grasslands, where dominant plant species benefit from herbivory through positive feedbacks between herbivores, plants and soil biota, and through preventing colonisation by later successional plants that produce poorer litter quality (redrawn from Bardgett & Wardle, 2003).

It is well known that plants allocate large proportions of their assimilated C to root exudation, which may stimulate the growth and activity of heterotrophic microbes in the rhizosphere. A number of studies have shown that defoliation of grassland plants actually leads to short-term pulse of root exudation, which has the effect of stimulating microbial activity in the rhizosphere (Mawdsley & Bardgett, 1997; Guitian & Bardgett, 2000; Paterson & Sim, 1999; Murray et al., 2004). In turn, this has been shown to also increase population densities of animals in the root zone that feed on microbes, such as bacterial and fungal-feeding nematodes (Mikola et al., 2001, Hokka et al., 2004). This is of relevance to sustainable agriculture because such positive effects of defoliation on rhizosphere organisms can feedback positively to the plant through enhanced soil N availability (Bardgett et al., 1998). This was recently demonstrated by Hamilton & Frank (2001) who showed that simulated defoliation of the grazing tolerant grass Poa pratensis led to increased leaf photosynthesis and root exudation of recently assimilated ^{13}C, which stimulated microbial biomass in the root zone. This in turn increased soil N availability and plant N acquisition, which ultimately benefited plant growth. It was proposed that such mechanisms could explain, in part, the compensatory response of grasses to grazing in high fertility grasslands (Hamilton & Frank, 2001) (Figure 2). There does appear to be much inter-species variability in the response of plants to defoliation (Guitian & Bardgett, 2000, Hokka et al., 2004). Also, responses are likely to vary with growth stage of the plant. However, what these findings collectively suggest is that physiological responses of plants to grazing have the potential to stimulate rhizosphere processes that ultimately feedback positively on plant nutrition and plant productivity (Bardgett & Wardle, 2003). Understanding the nature of such feedbacks between soil microbes, plants, and their consumers is clearly of high significance for sustainable grassland management, since nutrient cycling processes in unfertilised grasslands will be in part reliant on them. The challenge is thus to encourage such positive feedback mechanisms in sustainable grassland systems.

Case 3: Managing grassland to promote linkages between soil and plant diversity

The final example that I wish to consider is the potential for soil biodiversity to influence vegetation diversity in grassland. A key aim of sustainable management is the promotion of botanical diversity, and recent studies point to an important role for soil organisms in achieving this goal. This is perhaps best illustrated by using recent studies on mycorrhizal fungi, which are well known to benefit the performance of their host plants. However, whilst mycorrhizal fungi associate with the majority of plants within any community, they confer different degrees of benefit on certain plant species, thereby directly influencing the structure of plant communities. One of the first studies to illustrate this experimentally was by Grime et al. (1987) who assembled diverse grassland communities in microcosms and allowed them to develop in the presence and absence of arbuscular mycorrhizal (AM) fungi. The presence of AM fungi led to a shift in plant community composition, reducing the dominance of grass in favour of several subordinate herb species that benefited most from AM fungal infection. The net effect of this was a significant increase in plant species diversity due to a relaxation of plant competitive interactions. Positive effects of AM fungi on plant diversity have also been shown to occur in the field, due again to the promotion of subordinate species in the community (Gange et al., 1993). Several mechanisms have been proposed to explain increases in plant diversity resulting from AM fungal infection. Allen (1991) suggested that mycorrhizal fungi might increase plant diversity due to spatial heterogeneity of fungal infectivity in field soil, allowing mycotrophic (i.e. species that depend on mycorrhizal associations) and non-mycotrophic species to coexist in patches of high and low inoculum. Alternatively, Grime et al. (1987) suggested that AM fungi increase plant diversity through inter-plant transfers of carbon and nutrients via hyphal links, which lead to more even distribution of resources within the plant community, reducing the ability of certain species to monopolise resources. The net effect of such nutrient distribution would be more equitable competitive interactions between plant species, promoting species coexistence and greater plant diversity. This view is also consistent with findings of Van der Heijden (2004) who found that AM fungi promoted seedling establishment in perennial grassland by integrating emerging seedlings into extensive hyphal networks and by supplying nutrients, especially P, to the seedlings. AM fungi therefore act as a symbiotic support system that promotes seedling establishment and reduces recruitment limitation in grassland (Van der Heijden, 2004).

Variations in AM fungal diversity are also likely to strongly affect plant diversity in grassland. This was tested by Van der Heijden et $al.$ (1998b), who manipulated the number of native AM species (1-to-14 species) in experimental grassland units containing 15 plant species. They found that plant diversity and productivity were positively related to AM fungal diversity. This effect was thought to be due to an increase in AM hyphal length in the more diverse treatments, enabling more efficient exploitation and partitioning of soil P reserves, thereby relaxing plant competition and increasing plant productivity.

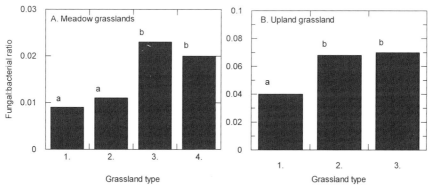

Figure 3 Effects of management intensity on fungal-to-bacterial biomass ratios measured using PLFA. (A). Gradient of declining management intensity on meadow grasslands in northern England, characterised by long-term reductions in fertiliser use and livestock density, classified as: (1) improved meadow; (2) very modified meadow; (3) slightly modified meadow, and (4) unmodified meadow. (B). Gradient of three upland grassland types of varying management intensity across ten sites in different biogeographic areas of the UK, from: (1) improved *Lolium - Cynosurus* grassland; (2) semi-improved *Festuca - Agrostis - Galium* grassland, *Holcus - Trifolium* sub-community, and; (3) unimproved *Festuca - Agrostis - Galium* grassland. Values are means and bars with the same letter are not significantly different (data from Bardgett & McAlister, 1999; and Grayston *et al.*, 2004).

What is clear from the above studies is that AM fungi have the potential to promote plant diversity in grassland, and also more efficient exploitation of nutrients within the plant-soil system. The challenge therefore is to develop management strategies that are specifically aimed at encouraging the growth of these fungi in grassland soil, with a view to enhancing plant diversity and nutrient exploitation. Recent literature shows that there is much scope to do this. In particular, studies of gradients of management intensity in temperate grasslands reveal that soils of traditionally managed, unfertilised grasslands have fungal-dominated food webs, with a high component of AM fungi, whereas intensive systems, characterised by high levels of fertiliser use, high grazing pressures, and reduced soil organic matter content, consistently have bacterial-dominated soil food webs (Yeates *et al.*, 1997, Bardgett & McAlister, 1999; Donnison *et al.*, 2000, Grayston *et al.*, 2001, 2004) (Figure 3). Furthermore, field manipulation studies reveal that seeding of particular plant species, for example legumes, can positively influence the growth of fungi in soils (Smith *et al.*, 2003), thereby increasing opportunities for plant-soil interactions that positively influence nutrient cycling and plant diversity in grassland. In sum, there is clearly much scope to manage grassland to promote plant-soil microbial interactions that are central to sustainable management.

Conclusions

This paper provides insights into fast developing areas of ecological science that offer scope for enhancing efficient cycling of nutrients in sustainable grassland systems. The work that I have presented has not yet been integrated into practical agriculture. However, what I have hopefully shown is that there is much scope to alter grassland management systems to optimise biotic interactions between plants, their grazers, and soil microbes, to increase the efficiency of nutrient cycling and plant nutrition in unfertilised, low input systems. Much work is still to be done, but the rapid development of this area of research suggests that integration of below-ground ecology into sustainable grassland management is not far away.

References

Allen, M. F. (1991). *The Ecology and Mycorrhizae*. Cambridge University Press, Cambridge.

Bardgett, R. D. (2005). *The Biology of Soil: A Community and Ecosystem Approach*. Oxford University Press.

Bardgett, R. D. & E. McAlister (1999). The measurement of soil fungal:bacterial biomass ratios as an indicator of ecosystem self-regulation in temperate grasslands. *Biology and Fertility of Soils*, 19, 282-290.

Bardgett, R. D. & D. A. Wardle (2003). Herbivore mediated linkages between above-ground and belowground communities. *Ecology*, 84, 2258-2268.

Bardgett, R. D., D. A. Wardle & G. W. Yeates (1998). Linking above-ground and below-ground food webs: how plant responses to foliar herbivory influence soil organisms. *Soil Biology and Biochemistry*, 30, 1867-1878.

Bardgett, R. D., T. Streeter & R. Bol (2003). Soil microbes compete effectively with plants for organic nitrogen inputs to temperate grasslands. *Ecology*, 84, 1277-1287.

Chapin, F.S., L. Moilanen & K. Kielland (1993). Preferential use of organic nitrogen for growth by a non-mycorrhizal arctic sedge. *Nature*, 361, 150-15.

De Mazancourt, C., M. Loreau & L. Abbadie (1999). Grazing optimization and nutrient cycling: potential impact of large herbivores in a savannah system. *Ecological Applications*, 9, 784-797.

Donnison, L. M., G. S. Griffith, J. Hedger, P. J. Hobbs & R. D. Bardgett (2000). Management influences on soil microbial communities and their function in botanically diverse haymeadows of northern England and Wales. *Soil Biology and Biochemistry*, 32, 253-263.

Gange, A. C., V. K. Brown & G. S. Sinclair (1993). Vesicular-arbuscular mycorrhizal fungi: a determinant of plant community structure in early succession. *Functional Ecology*, 7, 616-622.

Grayston S. J., C. D. Campbell, R. D. Bardgett, J. L. Mawdsley, C. D. Clegg, K. Ritz, B. S. Griffiths, J. S. Rodwell, S. J. Edwards, W. J. Davies and D. J. Elston (2004). Assessing shifts in soil microbial community structure across a range of grasslands of differing management intensity using CLPP, PLFA and community DNA techniques. *Applied Soil Ecology*, 25, 63-84.

Grayston, S. J, G. Griffiths, J. L. Mawdsley, C. Campbell & R. D. Bardgett (2001). Accounting for variability in soil microbial communities of temperate upland grasslands. *Soil Biology and Biochemistry*, 33, 533-551.

Grime, J. P., J. M. Mackey, S. H. Hillier & D. J Read (1987). Floristic diversity in a model system using experimental microcosms. *Nature*, 328, 420-422.

Guitian, R. & R. D. Bardgett (2000). Plant and soil microbial responses to defoliation in temperate semi-natural grassland. *Plant and Soil*, 220, 271-277.

Hamilton, E.W. & D. A. Frank (2001). Can plants stimulate soil microbes and their own nutrient supply? Evidence from a grazing tolerant grass. *Ecology*, 82, 2397-2402.

Henry, H.A.L. & R. L. Jefferies (2003). Plant amino acid uptake, soluble N turnover and microbial N capture in soils of a grazed Arctic salt marsh. *Journal of Ecology*, 91, 627-636.

Hodge, A., J. Stewart, D. Robinson, B. S. Griffiths & A. H. Fitter (1998). Root proliferation, soil fauna and plant nitrogen capture from nutrient-rich patches in soil. *New Phytologist*, 139, 479-494.

Hodge, A., J. Stewart, D. Robinson, B. S. Griffiths & A. H. Fitter (1999). Plant, soil fauna and microbial responses to N-rich organic patches of contrasting temporal availability. *Soil Biology and Biochemistry*, 31, 1517-1530.

Hokka, V., J. Mikola, M., Vestberg & H. Setälä (2004). Interactive effects of defoliation and an AM fungus on plants and soil organisms in experimental legume-grass communities. *Oikos*, 106, 73-84.

Jones D.L. & K. Kielland (2002). Soil amino acid turnover dominates the nitrogen flux in permafrost-dominated taiga forest soils. *Soil Biology and Biochemistry*, 34, 209-219.

Jones, D.L., J. R. Healey, V. B. Willett, J. F. Farrar & A. Hodge (2005). Dissolved organic nitrogen uptake by plants—an important N uptake pathway? *Soil Biology and Biochemistry*, 37, 413-423

Kielland, K. (1994). Amino acid absorption by arctic plants: implication for plant nutrition and nitrogen cycling. *Ecology*, 75, 2373-2383.

Lipson, D.A. & R. K. Monson (1998). Plant-microbe competition for soil amino acids in the alpine tundra: effects of freeze-thaw and dry-rewet events. *Oecologia*, 113, 406-414.

Mawdsley, J. L. & R. D. Bardgett (1997). Continuous defoliation of perennial ryegrass (*Lolium perenne*) and white clover (*Trifolium repens*) and associated changes in the microbial population of an upland grassland soil. *Biology and Fertility of Soils*, 24, 52-58.

McKane, R.B., L. C. Johnson, G. R. Shaver, K. J. Nadelhoffer, E. B. Rastetterk, B. Fry, A. E. Giblin, K. Kielland, B. L. Kwiatkowski, J. A. Laundre & G. Murray (2002). Resource-based niches provide a basis for plant species diversity and dominance in arctic tundra. *Nature*, 413, 68-71

McNaughton, S. J. (1985). Ecology of a grazing system: the Serengeti. *Ecological Monographs*, 55, 259-294.

Mikola, J., G. W. Yeates, G. M. Barker, D. A. Wardle & K. I. Bonner (2001). Effects of defoliation intensity on soil food-web properties in an experimental grassland community. *Oikos*, 92, 333-343.

Miller, A. E. & W. D. Bowman (2002). Variation in nitrogen-15 natural abundance and nitrogen uptake traits among co-occurring alpine species: do species partition by nitrogen form? *Oecologia*, 130, 609-616.

Miller, A. E. & W. D. Bowman (2003). Alpine plants show species-level differences in the uptake of organic and inorganic nitrogen. *Plant and Soil*, 250, 283-292.

Murray, P. J., N. Ostle, C. Kenny & H. Grant (2004). Effect of defoliation on patterns of carbon exudation from *Agrostis capillaris*. *Journal of Plant Nutrition and Soil Science*, 167, 487-493.

Näsholm, T., K. Huss-Danell & P. Högberg (2000). Uptake of organic nitrogen in the field by four agriculturally important plant species. *Ecology*, 81, 1155-1161.

Owen, A. G. & D. L. Jones. (2001). Competition for amino acids between wheat roots and rhizosphere microorganisms and the role of amino acids in plant N acquisition. *Soil Biology and Biochemistry*, 33, 651-657.

Paterson, E. & A. Sim (1999). Rhizodeposition and C-partitioning of *Lolium perenne* in axenic culture by nitrogen supply and defoliation. *Plant and Soil*, 216, 155-164.

Raab, T. K., D. A. Lipson & R. K. Monson (1999). Soil amino acid utilization among species of the cyperaceae: plant and soil processes. *Ecology*, 80, 2408-2419.

Schimel, J. P. & F. S. Chapin (1996). Tundra plant uptake of amino acid and NH_4^+ nitrogen in situ: plants compete well for amino acid N. *Ecology*, 77, 2142-2147.

Smith, R. S, R. S. Shiel, R. D. Bardgett, D. Millward, P. Corkhill, G. Rolph & P. J. Hobbs (2003). Diversification management of meadow grassland: plant species diversity and functional traits associated with change in meadow vegetation and soil microbial communities. *Journal of Applied Ecology*, 40, 51-64.

Streeter, T. C., R. Bol & R. D. Bardgett (2000). Amino acids as a nitrogen source in temperate upland grasslands: the use of dual labelled (^{13}C, ^{15}N) glycine to test for direct uptake by dominant grasses. *Rapid Communications in Mass Spectrometry*, 14, 1351-1355.

Turner, B. L., R. Baxter, N. Mahieu, S. Sjögersten & B. Whitton (2004). Phosphorus compounds in subarctic Fennoscandian soils at the mountain birch (*Betula pubesecens*)-tundra ecotone. *Soil Biology and Biochemistry*, 36, 815-823.

Van der Heijden, M. G. A. (2004). Arbuscular mycorrhizal fungi as support systems for seedling establishment in grassland. *Ecology Letters*, 7, 293-303.

Van der Heijden, M.G.A., J. N. Klironomos, M. Ursic, P. Moutoglis, R. Streitwolf-Engel, R. Boller, A. Wiemken & I. R. Sanders (1998b). Mycorrhizal fungal diversity determines plant biodiversity, ecosystem variability and productivity. *Nature*, 396, 69-72.

Weigelt, A., R. Bol & R. D. Bardgett (2005). Preferential uptake of soil nitrogen forms by grassland plant species. *Oecologia*, 142, 627-635.

Yeates, G.W., R. D. Bardgett, R. Cook, P. J. Hobbs, P. J. Bowling & J. F. Potter (1997a). Faunal and microbial diversity in three Welsh grassland soils under conventional and organic management regimes. *Journal of Applied Ecology*, 34, 453-471.

Section 1

Soil biology and nutrient turnover

Benomyl effects on plant productivity through arbuscular mycorrhiza restriction in a Greek upland grassland

M. Orfanoudakis[1], A.P. Mamolos[2], F. Karanika[2] and D.S. Veresoglou[2]
Aristotle University of Thessaloniki, [1]Aristotle University of Thessaloniki, School of Forestry & Natural Environment, Laboratory of Forest Soils, [2]School of Agriculture, Laboratory of Ecology & Environmental Protection, 541 24 Thessaloniki, Greece, Email: mamolos@agro.auth.gr

Keywords: arbuscular mycorrhiza, grassland, northern Greece

Introduction Interactions between plants and microbes are important for plant community structure. Many plants establish symbioses with arbuscular mycorrhizal (AM) fungi, which play a central role in soil fertility, plant nutrition and the maintenance of stability and biodiversity within plant communities by improving uptake of nutrients and water. AM fungi can improve growth/performance in a variety of plant species by influencing intra- and interspecific competition of neighbouring plants and thus regulate coexistence and diversity in mixed communities. The aim was to study AMF effects on plant productivity and diversity in Greek upland grasslands.

Materials and methods The experimental site (40°48′N, 21°23′E at 1340 m), on a Typic Xerothrent soil (average depth <35cm), faced south with 805 mm annual precipitation. Four blocks of two treatments (control, benomyl at 2.5 g benlate (DuPont)/m^2) were randomly arranged in 4.25 x 4 m plots. Above-ground vegetation was harvested from two 50 x 50 cm quadrats randomly selected in each plot every 14 days (starting on 29 April) and dried at 75° C for 48h. Undisturbed root samples of *Agrostis capilaris*, *Phleum pratense* and *Plantago lanceolata* from each of the blocks were collected with steel cylinders (10 cm x 20 cm long) on 7 July. 50g soil collected from the 0-15cm soil layer was mixed and AMF spores counted with a dissecting stereoscope at 160x magnification. Roots from each plant were stained with methyl blue, following Koske & Gemma (1989). The extent of colonisation was assessed by gridline intersection. Data were statistically analysed using ANOVAs.

Results AM colonisation of *Agrostis*, *Phleum* and *Plantago* was significantly reduced after benomyl application, while spore numbers were not (Figure 1). Aboveground biomass production tended to be reduced with benomyl applications for plant community, grasses and forbs and individual species.

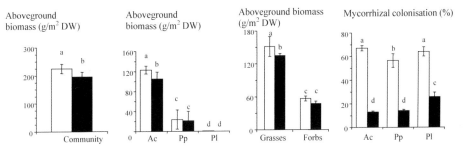

Figure 1 Effects of benomyl (mean ± 1 SE) on above ground biomass (Ac = *Agrostis*, Pp = *Phleum* and Pl = *Plantago*), and mycorrhizal colonisation: different letters = difference at $p = 0.05$

Conclusions Herbaceous grasslands in northern Greece, at altitudes >1000 m, consist of perennial C$_3$ species and are dominated by grasses and limited by water, N and P (Mamolos *et al.*, 2005). Benomyl application a) reduced colonisation of AM of *Agrostis capilaris*, *Phleum pratense* and *Plantago lanceolata*, b) tended to reduce aboveground productivity by restricting growth of dominant species, but c) did not affect AMF spore abundance. The initial data indicate that AM associations can influence productivity and plant diversity in such grasslands.

References
Koske R. E. & J. N. Gemma (1989). A modified procedure for staining roots to detect VA mycorrhizas. *Mycological Research*, 92, 486-505.
Mamolos A. P., C. V. Vasilikos & D. S. Veresoglou (2005). Vegetation in contrasting soil water sites of upland herbaceous grasslands and N:P ratios as indicators of nutrient limitation. *Plant & Soil* (in press).

The influence of burning on soil microbial biomass and activity along the Boro route in the Okavango delta of Botswana.

T. Mubyana-John and A. Banda

Department of Biological Sciences, University of Botswana, P/B 0022, Gaborone, Botswana, Email: mubyanat@mopipi.ub.bw

Keywords: fire, microbial activity, fungi, Okavango

Introduction The Okavango Delta, a protected area in northeastern Botswana because of its annual flooding pattern, is the main source of water in an otherwise arid environment with a high diversity of plants and animals and forms the main tourist area in the region. However, the area is under threat from range fires. The objective of this study was to assess the influence of fire on soil microbial activity, biomass C, fungal population and diversity, and some soil properties along the Boro route of the Okavango Delta (Botswana).

Materials and methods The influence of fire was assessed in grassland and Mopane woodland plots burnt between 2000 and 2003 and their adjacent unburnt plots in the 2003-04 flood and dry seasons. Soil microbial characteristics studied included soil dehydrogenase enzyme activity, fungal/bacteria biomass C ratio, soil fungal populations and diversity (Baisset & Parkinson, 1980). Soil physico-chemical properties studied included total nitrogen, pH, organic matter content and texture (Anderson & Ingram, 1993).

Results Fungal population and diversity was significantly higher in burnt plots. *Aspergillus* and *Penicillium* spp were the most dominant isolated genera. Fungi such as *Acroconidiella troposoli, Aspergillus candidus, Cochliobolus lunatus, Drechslera* sp, *Exophiala jeanselmei, Penicillium compactum*, and *Chrysosporium merdarium*, were only found in the burnt plots. *Fusarium* spp. dominated in the burnt plots, while *Aspergillus* spp. were low in the burnt plots. In contrast, bacterial biomass was not significantly different between burnt and unburnt plots. Soil dehydrogenase activity was higher in the burnt than in unburnt plots. The effect of burning on soil organic matter content depended on vegetation cover. Grassland plots showed significant decreases in organic matter after burning while effects in the Mopane woodland plots were minimal. Burning did not have a significant effect on soil nitrogen, an aspect attributed to the nitrogen being of flood water origin in the Okavango Delta.

Table 1 Influence of burning on soil properties

Plot	Treatment	Enzyme µg TPF/g	% Total soil N	Soil texture
6(00)	Unburnt	98 ± 6 **	0.03	S
	Burnt	151 ± 7	0.34	S
6(01)	Unburnt	123 ± 17	0.02	S
	Burnt	112 ± 11	0.01	S
2(03)	Unburnt	157 ± 2 *	0.05 **	SL
	Burnt	198 ± 13	0.12	SL
3(03)	Unburnt	134 ± 7 ***	0.38 *	S
	Burnt	72 ± 2	0.35	S
4(03)	Unburnt	115 ± 2	0.31	SL
	Burnt	149 ± 14	0.30	SL
5(03)	Unburnt	137 ± 23	0.21	S
	Burnt	137 ± 4	0.20	S

*p<0.05; **p<0.01; ***p<0.001; blank: NS

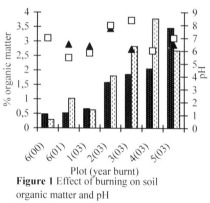

Figure 1 Effect of burning on soil organic matter and pH

■ Burnt ▢ Unburnt ▲ pH burnt ▢ pH unburnt

Conclusions These results indicate that in the Okavango Delta, burning increases fungal population, diversity and biomass. However, the influence on organic matter varies with the type of vegetation. The influence of burning on soil dehydrogenase activity, total nitrogen and pH were minimal.

References

Anderson, J. M. & J. S. I. Ingram (1993). Tropical soil biology and fertility: A handbook of methods. CAB International, UK

Baisset, J. & D. Parkinson (1980). Long-term effect of fire on the composition and activity of the soil microflora of sub alpine, coniferous forest. Canada Journal of Botany, 58, 1704-1721.

Estimating nitrogen fixation by pastures on a regional or continental scale

M. Unkovich
Soil and Land Systems, University of Adelaide, Roseworthy, SA 5371, Australia,
Email: murray.unkovich@ adelaide.edu.au

Keywords: model, productivity, medicago, chamaecytissus, trifolium, clover, legume

Introduction With fertiliser N inputs dramatically increasing in Australia in recent years (Angus, 2001), regional and continental scale estimates of biological nitrogen fixation (BNF) are now required for assessing the risks of terrestrial and surface water eutrophication, groundwater contamination, and gaseous N emissions.

Methods A simple model was developed for assessing annual BNF inputs into different pasture types in areas equivalent to local government districts, for the years 1983 – 1997. An existing dataset (Pearson *et al.,* 1997 (ATPD)) provided data on pasture types, legume species and content, and groundcover for some 562 local government areas of the Australian crop and pasture zone. Pastures were aggregated into 15 different types, representing ecological responses to the principal environments. The pasture types were allocated proportionally to the total pasture area (largest source of error) for each region based on the ATPD. Data from the literature and unpublished studies on pasture dry matter (DM) production and seasonal rainfall (annual for perennial based pastures and April-October for annual pastures) was used to construct linear regressions of rainfall *vs* dry matter production. When coupled with annual rainfall datasets, this was used to estimate annual DM production for each pasture type in each of 409 areas. Annual legume DM was calculated from the percentage legume for each pasture type given in the ATPD. A literature survey was similarly used to estimate N_2 fixation/tonne legume dry matter, including an estimate for "below-ground N" (Unkovich & Pate, 2000).

Results Nitrogen fixed/t legume DM ranged from 22.7 kg/t for annual Medicago and most perennial legume based pastures, to 26.2 kg/t for annual Trifolium pastures, both include estimated fixed N in below-ground root and nodule fractions. Total N fixation in all pastures was estimated as 3.9 Mt in 1994, a low rainfall year, to 5.0 Mt in a wetter year (1992). 'Pasture types' in Table 1 reflect the pasture base, not necessarily principal legume components. Annual *Trifolium* pastures were estimated to contribute *ca* 1,339 kt N (23 million ha), and all pastures 4.6 Mt N (92 million ha). In the same year, N fertiliser use in Australia was *ca* 800 kt, rising to 1100 kt in 2000.

Table 1 Total pasture areas, total N fixed, and averaged N fixation/ha for 1996

Pasture type	ha x 10^3	kt N fixed	kg N fixed/ha
Tagasaste	49	3.8	77.4
Lpuinus cosentinii	428	58.3	136.4
Fertilised native annual	593	3.9	6.6
Native annual	2646	1.5	0.6
Native perennial	10508	154	14.7
Fertilised native perennial	1193	10.1	8.5
Oversown perennial	5228	49.7	9.5
Tropical grass	3161	542.9	171.7
Naturalised medic	3601	40.2	11.2
Annual grass	1114	7.6	6.8
Madicago sativa	5739	673.8	117.4
Perennial grass	10364	961.4	92.8
Perennial legume	1224	199.4	162.9
Annual *Medicago*	22462	453.5	20.2
Annual *Trifolium*	23438	1338.9	57.1
TOTAL	91748	4499	

Conclusion Pasture legumes are estimated to have contributed 80% of the N input into Australian agriculture in 1996. McLaughlin *et al.* (1992) estimated N fixation in pastures to be 1.5 Mt (1987-88), and Jenkinson (1990) gives global biological fixation to be 140 Mt. On the basis of the present data, continental scale estimates for biological N fixation may need to be revised upwards.

References

Angus, J. (2001). Nitrogen supply and demand in Australian agriculture. *Australian Journal of Agricultural Research,* 41, 277-288.

McLaughlin, M., I. Fillery & A. Till (1992). Operation of the phosphorus, sulfur and nitrogen cycles. *In:* R. Gifford & M. Barson (eds). *Australia's Renewable Resources: Sustainability and Global Change'.* AGPS, Canberra , pp. 67-116.

Pearson C. J., R. Brown, W. Collins, K. Archer, M. C. Wood, C. Petersen & B. Bootle (1997). An Australian temperate pastures database. *Australian Journal of Agricultural Research,* 48, 453-466.

Unkovich, M. & J. Pate (2000). An appraisal of recent field measurements of symbiotic N_2 fixation by annual legumes. *Field Crops Research,* 211, 211-228.

Cycling of N and P in grass-alone (*Brachiaria*) and mixed grass/legume (*Brachiaria/Desmodium ovalifolium*) grazed pastures in the Atlantic forest region of Brazil

R.M. Boddey[1], R.M. Tarré[1], R. Macedo[2], C. de P. Rezende[3], J.M. Pereira[3], B.J.R. Alves[1] and S. Urquiaga[1]
[1]*Embrapa Agrobiologia, C. P. 74.505, Seropédica, 23851-970, RJ, Brazil, Email: bob@cnpab.embrapa.br,*
[2]*Núcleo Centro-Oeste da Embrapa Gado de Leite, Embrapa CNPAF, C.P. 179, Goiânia, 75375-000, GO, Brazil,*
[3]*CEPLAC, Estação de Zootecnia do Extremo Sul da Bahia, BR 101, Km 754, Itabela, BA, Brazil*

Keywords: Brachiaria pastures, *Desmodium ovalifolium*, nitrogen, nutrient cycling, phosphorus

Introduction There are estimated to be > 80 M ha of *Brachiaria* pastures in the tropical regions of Brazil. When continuously grazed with only modest fertiliser inputs (and no N) these pastures can maintain reasonable cattle weight gains (>200 kg LWG/ha per year). However, without fertiliser and when overgrazed, LWGs fall rapidly to low levels. Recent studies have shown that N and P deficiency are the most important factors limiting productivity. The objective of this study was to study fluxes of N and P in the pasture system in order to understand resilience to poor management and minimum nutrient requirements to guarantee their sustainability.

Materials and methods The experiment which was installed in 1987 at the CEPLAC field station in the Atlantic forest region south of Bahia (16°39'S, 39°30'W, mean annual rainfall 1,300 mm, no marked dry season, temperature 19 - 29°C). There were three stocking rates (2, 3 and 4 head/ha) for both grass-alone and of *B. humidicola* and a mixed sward with the forage legume *Desmodium ovalifolium* (see Rezende *et al.*, 1999). Throughout 1995, rates of deposition and disappearance (decomposition) were evaluated every 28 days as described by Rezende *et al.* (1999). All litter samples were analysed for N (semi-micro Kjeldahl), and P (colorimetry after perchloric/nitric acid digestion). Animal consumption was evaluated on two occasions (May and November) using chromic oxide as an external indicator (Raymond & Minson, 1955) and cattle fitted with oesophageal fistulae for sampling. Fistula and faecal samples were analysed for N and P as above.

Results Only the results of the highest and lowest stocking rates are shown (Table 1). In the absence of the legume the proportion of N and P consumed were 36 and 63 %, and 58 and 80 %, for the lower and higher stocking rates, respectively. With the legume in the sward these proportions were 25 and 56 %, and 49 and 72 %, respectively. The consumption of legume was highest at the highest stocking rate. Although the benefit to animal weight gain (an increase from 505 and 555 kg LWG/ha per year) was modest, N recycled though the litter pathway was increased by almost 50 %. The cattle obtained ~3 kg P/head per year from available salt licks, but even with this, apparent P utilisation rates were very high at the highest stocking rate.

Table 1 Annual fluxes of N and P through consumed and unconsumed forage of a grass-alone (*B. humidicola*) and mixed legume-grass (*B. humidicola/D. ovalifolium*) pasture in the Atlantic forest region of the south of Bahia, Brazil, grazed at two contrasting stocking rates by Zebu cattle

Sward/	Deposited Litter			Forage Consumption			Nutrient deposited in:				Exported in	
							Urine		Faeces		LWG	
Stocking Rate	DM	N	P	DM	N	P	N	P	N	P	N	P
	Mg/ha	kg/ha		Mg/ha	---------- kg/ha ----------							
Grass-alone /SR 2	29.7	170	12.7	7.8	95.9	17.6	58.0	0.2	37.9	14.9	9.0	2.5
Grass-alone /SR 4	21.3	105	9.2	13.6	178.6	36.2	112.6	0.3	66.0	32.5	12.6	3.4
Mixed sward /SR 2	33.1	325	13.3	6.3	108.2	12.9	65.1	0.1	43.1	10.3	9.1	2.5
Mixed sward /SR 4	23.6	149	11.5	13.0	186.6	29.2	92.8	0.3	93.8	25.1	13.9	3.8

Conclusions While the introduction of a forage legume had little direct effect on cattle productivity, the increased deposition of N in litter in the pasture would allow higher stocking rates for longer periods without the necessity of N fertiliser. At high stocking rates consumption of P was high showing the necessity of regular P fertilisation for sustained animal production.

References
Raymond, W. F. & D. J. Minson (1955). The use of chromic oxide for estimating the faecal production of grazing animal. *Journal of the British Grassland Society*, 10, 282-96.
Rezende C. de P., R. B. Cantarutti, J. M. Braga, J. A. Gomide, J. M. Pereira, E. Ferreira, R. M. Tarré, R. Macedo, B. J. R. Alves, S. Urquiaga, G. Cadisch, K. E. Giller & R. M. Boddey (1999). Litter deposition and disappearance in *Brachiaria* pastures in the Atlantic forest region of the South of Bahia, Brazil. *Nutrient Cycling in Agroecosystems*, 54, 99-112.

40 years of studies on the relationships between grass species, N turnover and nutrient cycling in the Lamto reserve in the Ivory Coast (Côte d'Ivoire)

L. Abbadie[1] and J.C. Lata[2]

[1]ENS Labo. d'Ecologie, 46 rue d'Ulm, 75230 Paris, cedex 05, France, Email: abbadie@biologie.ens.fr, [2]Labo d'ESE, Bât 362 Université Paris-Sud 91405, Orsay cedex, France

Keywords: *Andropogoneae*, competition, nitrification inhibition, nitrogen conservation, plant adaptation

Introduction The Lamto Station, dominated by grass savannas, was created in 1963. Among other problems, the relationships between savanna grasses (mainly *Andropogoneae* supertribe) and nutrient cycling, mainly nitrogen (N), have been intensively studied. Such grass systems are of major interest. Savannas represent 25% of terrestrial biomes and are second to tropical forests in the contribution to terrestrial primary production and are predominant in African social and economic environments. The *Andropogoneae* grasses are of particular interest for pastures. Second, savannas are generally extremely nutrient-poor, especially for N, which often limits productivity. Third, little is known about possible controls of grasses on N processes (e.g. nitrification) which could provide plants with potential advantages in competing for N, and induce changes in system N balance. Finally, these areas are considered to be non-emitting for NO and N_2O as a result of extremely low nitrification.

Results For over 40 years, grass cartography, systems, reproduction and genetic variability have been studied in the Lamto savannas. Distinct high (HN) and low (LN) nitrification sites have been discovered. Nitrification was positively (HN) or negatively (LN) correlated with root densities of distinct grass populations of the same species (Figure 1, Lata *et al.*, 2000). *In situ* plots showed that populations directly controlled nitrification (Lata *et al.*, 2004) and were adapted to different N forms: NH_4^+ (LN) or NO_3^- (HN). In LN sites, the denitrification level was 10-fold less than HN sites. In competition experiments, LN grass out competed HN grass. Isotopic fractionation (^{15}N) (Abbadie *et al.*, 1992) showed that grasses used N mainly from their own root system. Recent experiments showed that other controls were exerted on N fluxes: climate (rainfall), trees or termite mounds.

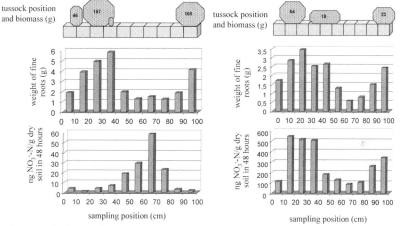

Figure 1 Transects in the LN site (left) and the HN site (right) at 0-10 cm depth. From top to bottom: tussock biomass, total fine roots weight and nitrifying enzyme activity

Conclusions An ability to inhibit nitrification, either through inhibition or superior plant competitiveness compared with microorganisms, could give a strong competitive advantage to LN plants. Our results suggest that grass species have important consequences for N cycling at population scales. For individual plants, N turnover is rapid and tightly controlled. Grasses seem well adapted to low N levels typical of the savanna ecosystems and, with other environmental patterns, create a high degree of 'patchiness' of nutrient resources at landscape scales.

References

Abbadie, L., A. Mariotti & J. C. Menaut (1992). Independence of savanna grasses from soil organic matter for their nitrogen supply. *Ecology*, 73, 608-613.

Lata, J. C., K. Guillaume, V. Degrange, L. Abbadie & R. Lensi (2000) Relationships between root density of the African grass *Hyparrhenia diplandra* and nitrification at the decimetric scale: an inhibition–stimulation balance hypothesis. *Proceedings of the Royal Society of London. Series B, Biological Sciences*, 267, 1–6.

Lata, J. C., V. Degrange, X. Raynaud, P.-A. Maron, R. Lensi & L. Abbadie (2004). Grass populations control nitrification in savanna soils. *Functional Ecology*, 18, 605-611.

The addition and cessation of inorganic fertiliser amendments in long-term managed grasslands: impacts on above and below-ground communities

C.D. Clegg[1], P.J. Murray[1], R. Cook[2] and T. Tallec[3]

[1]Institute of Grassland and Environmental Research, North Wyke Research Station, Okehampton, Devon EX20 2SB, U.K., Email: christopher.clegg@bbsrc.ac.uk, [2]Institute of Grassland and Environmental Research, Aberystwyth Research Centre, Aberystwyth, Ceredigion, SY23 3EB, UK, [3]Unité Mixte de Recherche INRA UCBN, Physiologie et Biochimie Vegetales, IRBA, Universite, 14032 Caen Cedex, France

Keywords: bacteria, fungi, plants, collembola, mites, nematodes, soil ecology

Introduction In recent times, land use in the United Kingdom has undergone considerable changes because of social and economic pressures, leading to a fine balance between the demands of highly productive intensive systems and practices which are perceived to be more environmentally acceptable. Plant productivity is governed by the supply of nutrients from the soil, which in turn is dependent on the dynamics of organic matter decomposition driven by soil micro-, meso- and macro fauna. Considerable information is available concerning the impact of inorganic fertiliser additions on communities of macro-fauna and flora, but the effects on specific microbial communities in soils are less clear. The effects of withholding inorganic nitrogen (N) are much less studied. The present study investigated the impact on plant and soil communities of either adding or with-holding N from long-term managed plots.

Materials and methods Soils (poorly drained silty clay loam of the Hallsworth series) were sampled from the Rowden long-term experimental site at the Institute of Grassland and Environmental Research, North Wyke Research Station, Devon in S.W. England. These were grazed swards that had been under long-term pasture for the previous 50 years and under experimental management for 15 years. Nitrogen fertiliser applications of 0 to 200 kg N/ha had been applied during this period and in the last two years additional treatments were introduced so that some previously unfertilised plots were treated with 200 kg N/ha and vice-versa. Botanical composition of the treatments was assessed visually and scored on a percentage scale. Soil samples were taken from each plot and analysed for the following characteristics. Soil microbial communities were determined through both phospholipid fatty acid (PLFA) and DNA-molecular profiling. Total direct bacterial counts and fungal hyphal lengths were determined microscopically. Soil nematode communities were described by trophic composition and by ecological indices based on proportions of five groups on the coloniser-persister scale. To represent soil mesofauna, soil collembola and mite communities were assessed by extracting the animals from the soil using a Tullgren funnel apparatus, individuals were assigned to morphotypes. Microbial PLFA, nematodes, mesofauna and plant data were statistically analysed using PCA and microbial DNA-molecular profiles by PCO. To establish the weight of impact of either adding or withholding N fertiliser to respective plots numerical distances were determined on all groups using canonical variance analysis.

Results and discussion Plant community composition was significantly different between the long-term fertilised and unfertilised plots. The addition of N to previously unfertilised plots had a significant impact on botanical composition whereas withholding N from the long-term fertilised plots had no significant effect. Total bacterial counts (typically about 10^8/g soil) and bacterial PLFAs were generally higher in those plots which were unfertilised, and bacterial molecular profiles were different between all treatments. Fungal hyphal lengths were highly variable between treatments, typically ranging from about 7–17 m/g soil, and were not significantly different. However, the amount of fungal PLFA 18:2ω6 was highest in the long-term unfertilised and lowest in the long-term fertilised treatments. For total microbial PLFAs, withholding N had a slightly larger effect than adding N. Faunal community composition was significantly different between the two long-term treatments. Numbers and diversity indices of mites and nematodes showed little change in response to a shift in fertiliser regime, however, greater numbers of collembola were supported in those soils that currently received N in both long-term background treatments.

Conclusions Whilst confirming the findings of previous studies reporting the impact of N addition on plant, micro- and mesofauna community compositions, a significant finding in this study was the impact of withholding N fertiliser on the composition of soil communities. It is not clear whether the effects seen were a result of the direct impact of inorganic N management on specific communities or the consequences of a series of trophic interactions. This may have implications for organic conversion, sustainability and restoration of previously managed grasslands.

Acknowledgments IGER is supported by the Biotechnology and Biological Sciences Research Council.

Grassland management practices and the diversity of soil nematode communities

R. Cook[1], P.J. Murray[2] and K.A. Mizen[1]

[1]*Institute of Grassland and Environmental Research, Aberystwyth Research Centre, Aberystwyth, Ceredigion, SY23 3EB, UK,* [2]*Institute of Grassland and Environmental Research, North Wyke Research Station, Okehampton, Devon, EX20 2SB, UK, Email: phil.murray@bbsrc.ac.uk*

Keywords: plants, nematodes, soil ecology

Introduction Nematodes are numerically abundant in northern temperate grassland soils where, through their feeding on plants, soil microbes and each other as well as being a food resource, they contribute to soil functioning and affect plant soil interrelationships. Permanent plant cover and the consequent abundance of root tissue supply a year-round food supply. There are only limited data on the effects of root-feeding by nematodes on the growth and development of pasture plants but under some circumstances above-ground biomass may be reduced. Herbivory by specific nematode parasites may not only directly affect the host plant but also promote soil microbial activity and nutrient fluxes. Nematodes in other feeding groups interact directly with the microbial communities influencing soil processes, including decomposition and mineralisation.

Materials and methods At the upland field site of the Natural Environment Research Council's Soil Biodiversity programme, plots of an established *Festuca - Agrostis* pasture and of a newly sown monoculture of *Lolium perenne* were treated for 3 years with nitrogen (as NH_4NO_3 at 240 kg/ha per year in two doses) and lime (as $CaCO_3$ at 6 t/ha per year at the beginning of the growing season) (N+L), or with pesticide (chlorpyrifos, 1.5 kg a.i./ha after each herbage cut, starting in late June 1999). The herbage was cut and removed monthly from June to September from 1999 to 2001. These treatments were designed to produce contrasting degrees of diversity in communities of soil animals and microbes as part of a study of the relationships between grassland management practice and the diversity and functions of soil biological communities. This paper describes the responses of nematode communities described by trophic composition and by ecological indices based on life history and reproductive potential ranked as five groups (c-p1, colonisers to c-p5, persisters) from soil samples taken in October each year. Details of responses and differences (at $P<0.05$) are based upon the analyses of transformed data, means are for non-transformed data.

Results and discussion N+L treated plots had fewer nematodes (2×10^7 per m^2) than either untreated controls or pesticide treated plots (both 3×10^7) and there were fewer nematodes in reseeded than in main plots (1 and 4 $\times 10^7$, respectively). Main plot treatments had similar numbers of nematodes in all feeding groups in 1999 but thereafter N+L plots had fewer plant-hyphal and fungal feeders but more bacterial feeders and predators. Pesticide had no effect. The reduced abundance in response to reseeding was shown in 1999 and 2000 by groups feeding on plant roots, in all years by plant-hyphal feeders, in 1999 by fungal feeders and in 1999 and 2000 by omnivores. Communities in original untreated swards were predominantly plant and plant-hyphal feeders (together more than 60%) and similar proportions (*ca* 25% each) of fungal and bacterial feeders. The changes in abundance were reflected in changed community structures. Abundances of c-p groups 1 and 2 were affected by the main treatments: the N+L plots had more c-p1 nematodes than either untreated or pesticide treated plots in 2001, but fewer c-p 2 in 2000 and 2001. Abundances of c-p 3 and 4 were not affected. Reseeding had marked effects on all four groups: c-p 1 being more abundant in 1999 and the other groups less abundant than in original swards in at least some years. The community composition in terms of c-p groups was affected by these changes in abundance. Nematodes in the c-p 2 group were predominant in the original untreated swards (>50%), followed by c-p 3 (*ca*33%) and similar smaller proportions (*ca*15%) of c-p 1 and 4.

Conclusions The trauma of cultivation, reseeding and a less diverse plant community was expected to reduce nematode numbers. The observed changes in nematode abundance and community structure in response to N+L can be correlated with a wide range of biotic and abiotic co-variates. The major impacts appear to be that N+L plots had increased plant inputs, promoting bacterial at the expense of fungal food channels, and that increased evapotranspiration of larger plants reduced soil moisture restricting nematode activity and multiplication. In addition, soil chemical changes may have directly affected some nematode taxa. The observations demonstrate how complex influences and feedbacks between nematode and plant communities are mediated by variations in biotic and abiotic factors such as are affected by grassland management practices and global climate change.

Acknowledgments This work was conducted as part of the Natural Environment Research Council's Soil Biodiversity thematic programme. IGER is supported by the Biotechnology and Biological Sciences Research Council.

Study of characteristics of soil animals in halophilous plant communities of *Leymus chinensis* grasslands of northeast in China

X. Yin, Y. Zhang and W. Dong

College of Urban and Environmental Science, Northeast Normal University, Changchun, Jilin Province. P.R. China 130024, Email: yinxq773@nenu.edu.cn

Keywords: *Leymus chinensis* grassland, halophilous plant community, soil animals

Introduction We have researched soil animals in 8 types of halophilous plant communities of *Leymus chinensis* grasslands of Northeast China to characterise soil animal groups and explain the role and function of soil animals in grassland ecosystems (Richard & Roger, 1998) and provide a scientific basis for research to improve alkaline lands in these grasslands.

Methods We investigated within *Leymus chinensis* grasslands the plant communities *Aeluropus litoralis*(A), *Puccinellia tenuifolia* (B), *Suaeda hetroptera* (C), *Suaeda glauea* (D), *Suaeda corniculate* (E), *Kochia sieversiana* (F), *Artemisia anethifolia* (G)and *Puccinellia chinampoesis* (H). Each plant community was sampled randomly at 4 sites. Sample sizes were 50cm × 50cm (for large-scale soil animals) and 10cm ×10cm (for middle-small-scale soil animals). We sampled at depths of 0-5cm, 5-10cm, 10-15cm, 15-20cm and 20-30cm. Animals were separated from soil by handpicking, Tullgren funnels and Baremann funnels (Jun-ichi AOKI, 1973).

Results A total of 784 soil animals belonging to 50 groups of 3 phyla, 4 classes, 14 orders and 36 families were found. There were 25 groups of large-scale soil animals. There were 3 dominant and 7 frequent groups. The individual numbers of both dominant and frequent groups accounted for 93.8% of the total. There were 35 groups of middle-small-scale soil animals. There were 3 dominant and 12 frequent groups. The individual totals of dominant and frequent groups accounted for 90.9% of the total. All of these groups were the basic components of soil animal populations and they were distributed widely in *Leymus chinensis* (Yin Xiuqin, 2003). Trends in group numbers and individuals for both large-scale and middle-small-scale soil animals in different halophilous plant communities were different (Figure 1). In order to analyse the relationship between soil animals and different communities, we calculated the diversity index of large-scale and middle-small-scale soil animals in different halophilous plant communities. The results are shown in Table 1. Vertical changes of large-scale and middle-small-scale soil animals in different halophilous communities were different.

Table 1 The diversity index of soil animals in different halophilous plant communities

Community No.	A	B	C	D	E	F	G	H
Large-scale	1.82	1.74	0.90	2.30	1.61	1.98	1.82	1.50
Middle-small-Scale	1.40	2.71	2.43	2.29	2.41	2.46	1.16	2.07

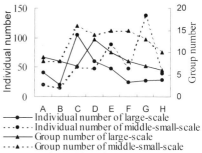

— Individual number of large-scale
···•··· Individual number of middle-small-scale
—▲— Group number of large-scale
···▲··· Group number of middle-small-scale

Figure 1 Number of individuals and groups of large-scale and middle-small-scale soil animals in different halophilous communities

Conclusions In the horizontal structure, for large-scale soil animals, the individual numbers of C was highest and B was lowest. However, the group numbers of D was highest and H was lowest. For middle-small-scale soil animals, the individual numbers of G was highest, B was lowest. While the group numbers of C was highest, A and B were the lowest. The group number and diversity index of soil animals were positively correlated. In 8 types of halophilous plant communities the numbers of groups and individuals of soil animals decreased with increasing soil depth.

References

Jun-ichi AOKI, (1973). The zoology of soil, Hokuryukan, Tokyo, 5-25.

Bardgett, R. & R. Cook (1998). Functional aspects of soil animal diversity in agricultural grassland. *Applied Soil Ecology,* 10, 263-276

Yin Xiuqin, Wang Haixia & Zhou Daowei (2003). Characteristics of soil animals communities in different agricultural ecosystem in the Songnen Grassland of China. *Acta Ecologica Sinica,* 23, 1071-1078

How soil properties affect egg development and larval longevity of a grassland insect pest - an empirically based model

S.N. Johnson[1], X. Zhang[2], J.W. Crawford[2], P.J. Gregory[1], S.C. Jarvis[3], P.J. Murray[3] and I.M. Young[2]

[1]School of Human & Environmental Sciences, Department of Soil Science, Whiteknights, University of Reading, Reading, RG6 6DW, UK, Email: S.N.Johnson@reading.ac.uk, [2]Scottish Informatics Mathematics Biology & Statistics (SIMBIOS) Centre, University of Abertay Dundee, Bell Street, Dundee, D1 1HG, UK, [3]Institute of Grassland and Environmental Research, North Wyke Research Station, Okehampton, Devon, EX20 2SB, UK

Keywords: desiccation, Fokker-Planck equation, temperature

Introduction The clover root weevil (*Sitona lepidus* Gyllenhal.) is a destructive pest of white clover in temperate grasslands. Adults lay thousands of eggs that give rise to soil-dwelling larvae that initially feed on the root nodules housing symbiotic N_2-fixing *Rhizobium* spp. bacteria. The period between egg hatch and consumption of root nodules by larvae is probably the most vulnerable part in the lifecycle, and if larvae do not locate roots relatively quickly they will die of starvation. In particular, the shells of eggs and the cuticles of emergent larvae are in constant physical contact with the external soil environment, so the nature of the soil is potentially critical for these life-stages. This study tested the effects of soil temperature, pH and moisture on egg development and subsequent longevity of unfed larvae to develop a mathematical model of these processes.

Materials and methods Freshly laid eggs were incubated in Eppendorf tubes containing soil under a range of different conditions; soil temperatures of 10°C, 15°C, 20°C and 25°C, soil pH 5, 7 and 9 and soil moisture contents of 0%, 10% and 25%. Tubes were examined daily to record when eggs hatched and the subsequent lifespan of the emergent unfed larvae. Statistical significances were determined using Cox proportional hazards regression models (Cox & Oates, 1984). Results were used to develop an individual-based model.

Results No eggs hatched in the dry soil (Figure 1). In damp soils, temperature had the greatest impact on egg development time with eggs developing faster but larval lifespan decreasing at higher temperatures. Soil pH also had a slight impact on egg development. There was no significant difference at soil moistures of 10% or 25% on egg development or larval lifespan. The Fokker-Planck equation (Jesperson *et al.*, 1999) was developed to model egg hatching and larval lifespan as a single process: simulations showed close fidelity with results (Figure 2).

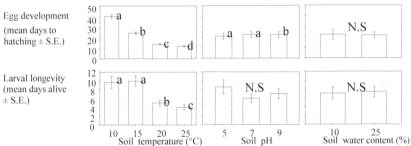

Figure 1 Effects of soils on egg development and larval longevity (Letters indicate significant differences)

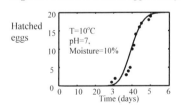

Figure 2 Example of model simulation (line) and experimental results (circles)

Conclusions Soil moisture is necessary for *S. lepidus* egg development. Temperature was the most significant factor tested with eggs developing faster, but larvae surviving less time at higher temperatures. An individual based model built on the Fokker-Planck equation permits these processes to be simulated as a single process.

References

Cox, D. R. & D. Oakes (1984). *Analysis of Survival Data*. Chapman & Hall, London.

Jespersen, S., R. Metzler & H. C. Fogedby (1999). Levy flights in external force fields: Langevin and fractional Fokker-Planck equations and their solutions. *Physical Review*, 59, 2736-2745.

Impact of root herbivory on grassland community structure: from landscape to microscale

P.J. Murray[1], R. Cook[2], L.A. Dawson[3], A.C. Gange[4], S.J. Grayston[3] and A.M. Treonis[3]

[1]Institute of Grassland and Environmental Research, North Wyke Research Station, Okehampton, Devon EX0 2SB, UK, Email: phil.murray@bbsrc.ac.uk, [2]Institute of Grassland and Environmental Research, Aberystwyth Research Centre, Aberystwyth, Ceredigion, SY23 3EB, UK, [3]The Macaulay Institute, Craigiebuckler, Aberdeen, AB15 8QH, [4]Department of Biology, Royal Holloway University of London, Egham, TW20 0EX, UK

Keywords: root herbivores, biodiversity, soil communities

Introduction Root herbivores are an important functional group in grassland ecosystems. Whilst there is a plethora of information on their impact as pests in productive grassland, few studies of their impact on biodiversity in upland grassland have been made. Root herbivores act in a number of ways, they reduce host plant biomass, alter root architecture, change root exudation patterns and increase water stress in the plant. Root herbivores may change above ground plant diversity, both through direct removal of plant species and through reduction in competitive ability of some species, through their feeding. In addition, we postulate that root herbivores affect soil microbial communities through changes in root exudation.

Materials and methods *Field Study.* As part of the UK NERC 'Soil Biodiversity' thematic programme, a randomised block experiment was established on an upland grassland (NVC classification U4d), at Sourhope in the Scottish Borders, UK. The experimental area was fenced to prevent grazing and comprised five replicate blocks each with six 20 x 12 m plots, marked out in late 1998. Plots in each block were allocated to one of five experimental treatments: 1: control (C), untreated (two plots); 2: lime (L) applied at a rate of 1.2 kg/m^2; 3: nitrogen (N) applied at a rate of 12 g/m^2; 4: lime and nitrogen (N+L) applied as above; and 5: pesticide applied as chlorpyrifos (Dursban 4: Dow Agrosciences) at 1.5 kg active ingredient per ha. The N and the lime were applied once per year at the beginning of the season. The plots are cut every three weeks from May into September (5 times per year). The pesticide was applied after every cut. Root herbivore populations in three of the treatments (C, N+L and pesticide) were sampled at three monthly intervals from June 1999 to March 2001: at the end of the experimental programme a detailed botanical survey was undertaken.

Microcosm studies. In order to determine feeding effects at a smaller scale, a number of microcosm experiments were undertaken to investigate the impact of root feeding on soil microbial community structure. *Lolium perenne*, *Trifolium repens* and *Agrostis capillaris* were grown individually in soil microcosms with larvae of *Tipula palidosa* added to half of the units. Microbial community structure under the plants was determined by Community Level Physiological Profiles (CLPP)

Results and discussion In the field study, populations of *Tipula paludosa* initially reached *ca* 120/m^2, but reduced to zero in the second year in the C and NL plots. However, populations of *Agrotis* spp. (cutworms) increased in these treatments, thus maintaining root herbivory in the system. The pesticide addition removed all large insects within three months. In addition to the more obvious impact of root herbivory (i.e. removal of plant tissue) root feeding had a number of other more subtle effects, both on the plant and on the other members of the soil community. The detachment of large quantities of root material puts severe pressure on the plants and demands re-allocation of resources for root maintenance and replacement. This is important in determining the fitness of individual plants with implications for plant diversity. Plant species richness was greater in the pesticide plots than in the C plots and this may be a consequence of the removal of the root herbivores, particularly as the overall numbers of the major above ground herbivores (slugs) were not reduced by the pesticide. The removal of large amounts of plant tissue by root herbivores increases the inputs to the detrital pool in the soil and provides an energy source for the soil micro-organisms. The soil microbial community is also influenced by changes in root exudation patterns, mediated by root feeding. In the microcosm experiments the presence of the larvae led to a change in the CLPP of the soil microbial community which was mainly due to differences in sugars, carboxylic and amino acid usage, suggesting larval herbivory increases the release of these compounds, which then selects for microorganisms capable of utilising these substrates.

Conclusions A wide range of effects of root herbivory in grassland systems was demonstrated. It should be noted that the presence of root feeders is common in grassland and their impacts should not be dismissed lightly.

Acknowledgments This work was funded by NERC. IGER is supported by BBSRC.

Analysis of the soil foodweb structure on organic- and conventional dairy farms

N. van Eekeren[1], F. Smeding[1] and A.J. Schouten[2]
[1]Louis Bolk Institute, Hoofdstraat 24, 3972 LA Driebergen, The Netherlands, Email: n.vaneekeren@louisbolk.nl,
[2]RIVM, P.O. Box 1, NL 3720 BA, Bilthoven, The Netherlands

Keywords: soil foodweb, soil biology, dairy farms, organic

Introduction The below ground biodiversity of soil organisms plays an important role in the functioning of the soil ecosystem, and consequently the efficiency of the above ground plant production on dairy farms. However, for farmers, soil biology remains a black box. It is difficult to interpret soil biology on dairy farms and to identify management measures to improve it. The objective of this study was to investigate if it is possible to classify the soil foodweb structure on a dairy farm in relation to management practices.

Materials and methods In an empirical study, soil foodweb structures of 49 dairy farms (including 10 organic farms) on sandy soils were related to farm and site characteristics. Soil life records included densities of bacteria, nematodes, enchytraeids, earthworms, springtails and mites. Sampling was done in 1999 under the Dutch national programme 'Biological Indicator for Soil Quality' (BISQ) (Schouten et al., 1999). Data were agglomerated in eleven trophic groups, and standardised to the group maximum. Samples were classified by means of TWINSPAN (Hill, 1979).

Results TWINSPAN clearly distinguished 5 main soil foodweb structures within 45 of the 49 dairy farms (Table 1). These soil foodweb structures, combined with farm characteristics, can be described as follows:
Type 1: Intensively managed organic farms with a high biomass of enchytraeids and earthworms;
Type 2: Organic- and extensive conventional farms with a high soil organic matter, high numbers of nematodes and a high biomass of bacteria;
Type 3: Extensive conventional farms with a high percentage of grassland and a high biomass of earthworms;
Type 4: Intensive conventional farms with more arable land and a reduced soil life;
Type 5: Intensive conventional farms with a reduced soil food web but high numbers of mites.

Table 1 Classification of 45 dairy farms in 5 types of soil foodweb structures

	Dairy farms			Site characteristics				Average quantities (see key)											
Type	No. organic	No. conv. ext.	No. conv. int.	% grassland	LU/ ha	SOM %	P-Al mg/100g	Bacteria	Nem hf	Nem pf	Nem bf	Nem mp	Enchytraeid	Earthworm	Mite top	Mite bhf	Mite pf	Miet mp
1	4	1	0	92	2.2	7.7	48	146	1	22	23	3	8	40	0	11	8	8
2	5	4	0	80	1.8	9.2	48	240	2	22	33	5	3	41	0	12	7	7
3	0	4	3	80	2.8	5.8	53	151	1	16	41	5	3	102	0	13	4	9
4	0	6	11	68	2.9	6.7	50	142	0	10	28	3	2	28	0	18	6	11
5	0	1	5	80	3.3	5.9	59	147	1	13	31	2	2	28	1	51	11	30

Key: LU=Livestock Unit, SOM=Soil Organic Matter; Bacteria (µg C/g soil), Nem=nematodes (number/g soil), hf=hyphen-feeding, pf=plantfeeding, bf=bacteria-feeding; mp=micropredator; Enchytraeid (g/m²); Earthworm (g/m²); Mite=mites and springtails (number/10cm²), top=arthropod predator, bhf=bacteria- & hyphen-feeding, pf=plant-feeding, mp=micro-predator.

Conclusions The results of this study demonstrate that the soil food web structure can be classified and is related to farm characteristics. The next step would be to determine the performance of the different structures in terms of soil functions and above ground production.

References
Hill, M. O. (1979) TWINSPAN - A FORTRAN program for arranging multivariate data in an ordered two-way table by classification of individuals and attributes. Cornell University Ithaca, N.Y., 90 pp.
Schouten A. J., J. Bloem, W. Didden, G. Jagers op Akkerhuis, H. Keidel & M. Rutgers (2002). Bodembiologische indicator. (1999). Ecologische kwaliteit van graslanden op zandgrond. RIVM report 607604003, 107 pp. (in Dutch, English Summary).

The effect of forage legumes on mineral nitrogen content in soil

M. Isolahti[1], A. Huuskonen[1], M. Tuori[2], O. Nissinen[3] and R. Nevalainen[1]

[1]MTT Agrifood Research Finland, North Ostrobothnia Research Station, FIN-92400 Ruukki, Finland, Email: mika.isolahti@mtt.fi, [2]Helsinki University, Department of Animal Science, FIN-00014 Helsinki University, Finland, [3]MTT Agrifood Research Finland, Lapland Research Station, FIN-96900 Saarenkylä, Finland

Keywords: legumes, nitrogen, soil

Introduction The cultivation of forage legumes is often suggested as a possibility to improve nitrogen (N) utilisation in farming. However, previous studies have indicated examples in which the cultivation of legumes such as white clover has led to accumulation of large amounts of N in soil (Adams & Pattison, 1985). In this study the potential risks of N leaching were estimated by determining amount of mineral N in the soil.

Materials and methods The effect of forage legumes to the mineral N (ammonium and nitrate) content in soil was studied in Finland in three different locations: one in Southern Finland in Helsinki (60°13') and two sites in Northern Finland Ruukki (65°40') and Rovaniemi (66°35'). The legume species were red clover (*Trifolium pratense* L.), white clover (*Trifolium repens* L.), alfalfa (*Medigaco sativa* L.), goats-rue (*Galega orientalis* Lam.) and common birds-foot-trefoil (*Lotus corniculatus* L.). Legumes were sown both as a monoculture and as a mixture with meadow fescue (*Festuca pratensis* Huds.). In addition, meadow fescue was grown as a pure culture with or without N fertiliser at 200 kg/ha. Soil samples were taken from three different layers: 0-33 cm, 33-66 cm and 66-100 cm and the experimental plots in three consecutive years both in the autumn after the growth period and in the spring soon after the melting of ground frost. After the cessation of third growth period the plant stands were ploughed and the soil samples were taken next autumn and the spring after.

Results The mineral N content in the soil samples in autumn was significantly higher in the pure forage legume stands than in the mixture stands or in pure meadow fescue stands (for instance white clover, Figure 1). This was a common feature to all study sites and years. The effect was observed for all legume species. However, there were no distinct differences between the pure legume stands. The plant stand with highest mineral N content in the soil varied between years and study sites. The highest mineral N contents were generally in the plots with high yield and presumably efficient N fixation. The mineral N content increased significantly from autumn to spring in every experiment site (Figure 1). In coarse soils the mineral N contents moved deeper more quickly than in clay soils. The leaching of N has probably taken place in every site, but the amount of mineralisation from soil organic matter has been greater than leaching. Nitrogen is mineralised during winter even in such harsh climatic conditions as in Finland (Bergström & Brink, 1986).

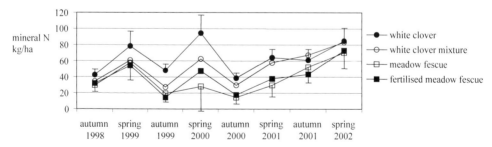

Figure 1 The amount of mineral nitrogen of meadow fescue, white clover and a mixed sward in the soil in Helsinki during 1998-2002. Estimates of confidence intervals (95%) are shown by error bars

Conclusions The cultivation of pure forage legumes will lead to accumulation of considerable amounts of organic N, which is vulnerable to leaching and denitrification when organic matter decomposes especially after cultivation of legume swards. One way to decrease the mineralisation of N is to leave tilling until late autumn.

References
Adams, J. A. & J. M. Pattison (1985). Nitrate leaching losses under a legume-based crop rotation in central Cantebury, New Zealand. *New Zealand Journal of Agricultural Research*, 28, 101 - 107.
Bergström, L. & N. Brink (1986). Effects of differentiated applications of fertiliser N on leaching losses and distribution of inorganic N in the soil. *Plant and Soil*, 93, 333-345.

Field experiments to help optimise nitrogen fixation by legumes on organic farms

A. Joynes[1], D.J. Hatch[1], A. Stone[1], S. Cuttle[1] and G. Goodlass[2]
[1]Institute of Grassland and Environmental Research, North Wyke Research Station, Okehampton, Devon, EX20 2SB,UK, Email adrian.joynes@bbsrc.ac.uk, [2]ADAS, High Mowthorpe, Duggleby, Malton, North Yorkshire, YO17 8BP, UK

Keywords: organic farming, legumes, nitrogen, fixation

Introduction During an organic rotation, the aim is to increase the nitrogen (N) content of the soil (and hence build up soil fertility) by recycling crop residues, applying manures/composts and from N fixed by legumes. IGER, with ADAS, Duchy College Cornwall and Abacus Organic Associates are developing improved guidance for organic farmers in the use of fertility-building crops. The main questions are: how to maximise N fixation and how to make the most efficient use of the fixed N? Available soil mineral N, which is generally thought to reduce N fixation, will be increased by manure applications, cutting/mulching and grazing. We describe an experiment to establish the extent to which animal and green manures can adversely affect N fixation. The results from the first year (2003) were reported recently (Hatch *et al.*, 2004). Here we summarise the findings from 2 years' results (2003-4) to show the changes that occurred after the legume crop became fully established.

Materials and methods A 2-year study was undertaken on a well drained, reddish gravelly loam soil from the Crediton Series at IGER, North Wyke in south west England. Forty eight paired plots (1.5 x 10m) were prepared for planting either with, or without, composted farmyard manure (FYM) at a rate of 170 kg N/ha in autumn 2002. The plots were randomised and one half was sown with red clover, while the other half was sown with ryegrass. While the ryegrass stayed relatively weed free, the clover plots contained a significant proportion of grass that germinated from the seed bed. The plots were cut 4 times during each growing season and the following treatments (6 replicates to each) were applied: A. Red clover (herbage cut and removed); B. Ryegrass control (herbage cut and removed); C. Red clover (herbage cut and returned to plot) and D. Ryegrass control (herbage cut and removed: herbage from treatment A spread on this plot). Herbage returned to plots was mulched by spreading and mowing and a second application of FYM was made in autumn 2003. Herbage samples from each cut were analysed for dry matter, total carbon and N. N fixation was estimated by subtracting the N yields from the ryegrass controls, from that found in the corresponding clover/grass plots. Annual N yields (Table 1) were obtained from the cumulative harvests of each growing season, by combining the data from the ± FYM treatments (as these were not found to be significantly different).

Table 1 Average N yields (kg N/ha per year) for treatments (combining the ± FYM data). Treatments (12 reps each) with different letter subscripts are significantly different (P < 0.05) within years

Treatments	N sources	Year 1 (2003)	Year 2 (2004)
B	N obtained from soil	120_c	91_c
D	N obtained from soil + mulch	146_b	135_b
D-B	N from mulch	26	44
A	N from soil + fixation	159_{ab}	212_a
A-B	N from fixation	39	121
C	N from soil + fixation + mulch	165_a	230_a
C-D	N fixation in presence of mulch	20	95
(A-B)-(C-D)	Loss in fixation caused by mulching	19	26

Results and discussion From the 170 kg N/ha applied as FYM each autumn, only 26 kg N/ha per year was recovered in the herbage and yield effects were only detected in the first harvests (data not shown). A trend for FYM to reduce N fixation was consistent, but not significant. Mulching increased the N yield of ryegrass (D) and clover/grass (C) in both years but the increase was only significant (P<0.05) in ryegrass. The benefit to clover was probably offset by a negative feed-back from added N in the mulch, leading to reduced fixation: a loss of 19 and 26 kg N/ha per year for years 1 and 2, respectively.

Acknowledgements This work was conducted as part of a Defra funded project (OFO316). IGER is supported by the Biotechnology and Biological Sciences Research Council.

References
Hatch, D. J., A. Joynes, A. Stone & G. Goodlass (2004). Soil fertility building crops in organic farming. Ramiran 2004-11th International Conference of the FAO ESCORENA network on Recycling of Agricultural, Municipal and Industrial Residues in Agriculture, Murcia, Spain, 6-9 October 2004, p33.

Effects of applied quantity of phosphorus fertiliser on phosphorus content in plant tissues of lucerne (*Medicago sativa*) and seed yield in North-western China

Y.W. Wang[1], J.G. Han[1], S.M. Fu[2] and Y. Zhong[2]
[1]Department of Grassland Science, China Agricultural University, Beijing, China 100094, Email: wangyunwen120@sohu.com, [2]Chengdu Daye International Investment Co. Ltd. Chengdu, China 610016

Keywords: lucerne, phosphorus fertiliser, tissue phosphorus concentration, seed yield

Introduction Phosphorus concentration in plant tissue can be a useful index of P deficiency in lucerne, P fertiliser recommendations and monitoring of effectiveness of current P fertiliser practices (Jacobsen & Surber, 1995). The objective of this study was to measure P concentration in different lucerne plant parts and seed yield in relation to P fertiliser application rates in order to improve recommendations for lucerne seed production.

Materials and methods The experiment was located in north-western China (39°37'N, 98°30'E, altitude 1480 m). The soil was an irrigated desert earth (Chinese soil classification), classified as silty clay soil with pH 8.5. Total soil P and available P content was 0.607 g/kg and 12.47 mg/kg, respectively. The lucerne was established with 60 cm row space and a seeding rate of 4.0 kg/ha. The experimental design was a randomised block with 3 replications of 7 P fertiliser application rates laid out in spring of 2001. Ten to fifteen plants of 15-20 cm high were taken per replicate in the regrowth period in spring of 2002; roots were 30-40 cm long. A sample of each plant part was taken by combining the plant parts of three replicates. P in each plant tissues and seeds were determined spectrophotometrically.

Table 1 The total P content of plant parts response to applied P fertiliser rates*

P$_2$O$_5$ kg/ha	Total P content (%DM)			
	Roots	Stems	Leaves	Seeds
0	0.09	0.23	0.34	0.45
60	0.20	0.27	0.42	0.56
120	0.19	0.30	0.40	0.63
180	0.21	0.31	0.41	0.64
240	0.23	0.28	0.40	0.62
300	0.20	0.29	0.40	0.65
360	0.23	0.30	0.42	0.67

* Effects of P fertiliser application on all alfalfa plant parts were significant (p<0.001)

Table 2 The equations of the relationship of total P content of plant parts to applied P fertiliser rates

Plant parts	Equation	Pr>F	R^2	S.E.
Roots	$Y=0.1138+0.000869X-1.65\times10^{-6}X^2$	0.05	0.77	0.028
Stems	$Y=0.2386+0.000530X-1.09\times10^{-6}X^2$	0.07	0.73	0.017
Leaves	$Y=0.3629+0.000405X-7.94\times10^{-7}X^2$	0.28	0.47	0.024
Seeds	$Y=0.4717+0.00134X-2.35\times10^{-6}X^2$	0.01	0.89	0.031

Results Addition of 60 to 360 kg P$_2$O$_5$/ha resulted in a relatively greater increase in P concentration for roots than other plant parts (Table 1). The responses of lucerne seed yield and seed yield components to P application treatments in two seasons were not significant (P>0.05), addition of 120 kg P$_2$O$_5$/ha treatment showed highest seed yield of 794.1 kg/ha and 757.1 kg/ha in both years (data not listed). The P contents in seeds, roots and stems were highly correlated to P application rate (r^2 = 0.89, 0.77, and 0.73, respectively) and to a lesser extent to P content in leaves (r^2 = 0.47). Quadratic regression equations for roots and seeds were significant (P≤0.05).

Conclusions The concentration of P in roots can be a good indicator of soil P fertility. The critical concentration of P in roots to determine P fertiliser recommendations for lucerne plant growth or seed production deserves further evaluation.

References
Jacobsen, J. S. & G. W. Surber (1995) Alfalfa/grass response to nitrogen and phosphorus application. *Communications in Soil Science and Plant Analysis*, 26, 1273-1282.

Cool-season grass response to increasing nitrogen fertiliser rates in Michigan

R.H. Leep, T.S. Dietz and D.H. Min
Michigan State University, Crop and Soil Sciences, East Lansing, MI 48824, USA, Email: leep@msu.edu

Keywords: nitrogen fertiliser, grass response

Introduction Nitrogen (N) fertility recommendations for cool-season grasses in the north central region of the USA have not been species specific. This broad recommendation assumes that all grasses have similar N demands, while seasonal growth patterns and dry matter yield of cool-season grass species vary. Nitrogen fertiliser costs have steadily increased, but recommendations are to be below optimal levels for economic return (Klausner *et al.*, 1998). A more specific N fertiliser recommendation may increase the producers' net income.

Materials and methods Five cultivars of four species ('Aries' diploid perennial ryegrass (PR) *(Lolium perenne)*, 'Quartet' tretaploid perennial ryegrass, 'Barolex' tall fescue *(Festuca arundinacea)*, 'Duo' festulolium *(Festuca x Lolium)*, and 'Sparta' orchard grass *(Dactylis glomerata)* were seeded in a split-block design with cultivar being the whole plot at East Lansing, MI., USA in 2001. Ammonium nitrate was applied in split applications (Table 1) to deliver 0,112, 224, 336, and 448 kg N/ha per year. Weeds were controlled as needed with Trimec (2,4-D, dicamba) herbicide. Yield, crude protein (CP), neutral detergent fibre (NDF), acid detergent fibre (ADF), tiller counts and visual ratings for ground cover and disease were collected annually.

Table 1 Nitrogen application (kg/ha) and timing

Rate	Timing
0	0
112	56 at new growth initiation, 56 after 1st cut
224	56 at new growth initiation , 56 after 1st cut, 56 after 2nd, 56 after 4th
336	112 at new growth initiation , 84 after 1st cut, 84 after 2nd, 56 after 4th
448	112 at new growth initiation , 112 after 1st cut, 112 after 2nd, 56 after 3rd, 56 after 4th

Results Three-year total yield responses by cultivar to N rates are presented in Figure 1. Both perennial ryegrass cultivars were winter killed and were reseeded in 2003 so yields of these are for 2002 and 2004 only. The greatest yield response from N application was observed in tall fescue. There was no gain in yield for any species above 224 kg N/ha. Tillering was increased by N applications in diploid PR and tall fescue. There was no change in ADF or NDF, but CP increased on average by 28 g/kg when N was applied at the lowest rate for all species. Leaf rust *(Puccinia brachypodii)* and Septoria leaf spot indices were decreased in all species by N application, but tall fescue had the greatest decrease from increased N.

Conclusions The importance of N is clearly demonstrated for all cultivars tested. While tall fescue was the most efficient in utilising N, all species optimised N use at 224 kg N/ha. Increased yield is often noted when N is applied, but increased CP and decreased occurrence of foliar disease should also be noted. The application of N encourages tiller initiation and this new growth is disease-free. Plants growing and initiating tillers more rapidly, due to N application, had less evidence of disease. While all grass species had the same optimal N level, the yield increase due to increasing the N rate from 122 to 224 kg/ha varied.

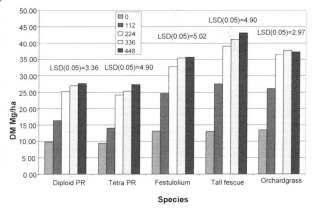

Figure 1 Yield response of pasture grasses to increasing nitrogen rates

References
Klausner, S. D., J. H. Cherney, R. F. Lucey & W. S. Reid. (1998). Nitrogen fertilization of grasses. Res. Ser. 98-1 College of Agric. and Life Sci. Cornell Univ., Ithaca, NY.

Within resting period seasonal soluble carbohydrate profiles of rotationally grazed elephant grass

L.P. Passos, M.C. Vidigal, I.G. Perry, F. Deresz and F.B. de Sousa
Embrapa Dairy Cattle, The National Dairy Cattle Research Centre, Juiz de Fora, 36016-210 MG, Brazil, Email: lpassos@cnpgl.embrapa.br

Keywords: elephantgrass, rotational grazing, soluble carbohydrates

Introduction Recent evidence indicates that carbohydrate-rich grazed herbage is effective for improving milk production (Trevaskis *et al.*, 2004). However, the dynamics of water-soluble carbohydrate (WSC) seasonal accumulation as related to forage availability remains unknown, especially during the growth stage of tropical forage grazing systems. The objective of the work was to verify the seasonal WSC profiles of rotationally grazed elephantgrass (*Pennisetum purpureum* Schum.), by making measurements within each 30-day resting period.

Materials and methods Rotationally grazed elephantgrass paddocks were randomly sampled every 10 days within each 30-day resting period during the 2001-2002 rainy season. Stem bases were harvested, oven-dried (65°C, 72 h), and their WSC contents determined by an autoclave extraction procedure described by Passos *et al.* (2003). In addition, five months were randomly chosen for estimation of forage availability by using the random plate method. The experiment was conducted as split-plots (months set as plots and days within resting periods as subplots), in a randomised block design with two replicates. The data were analysed by ANOVA and means among treatments compared by the Tukey test. For reasons of clarity, data are shown with the SDs.

Results Significantly higher WSC levels were observed only in the period immediately before growth ceased in June (Figure 1). However, consistent tendencies of stabilised WSC contents were verified in the middle of the rainy season, which were not apparently related to increased pasture on offer. Higher pasture availability, on the other hand, appeared to be preceded by steady WSC enhancement during the resting period.

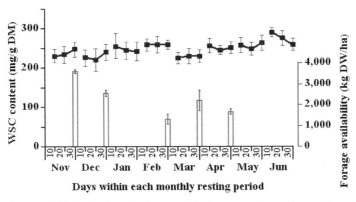

Figure 1 WSC levels of rotationally grazed elephantgrass within each monthly resting period

Conclusions The initial trends verified in the present study suggest that, within the resting period, rotationally grazed elephantgrass WSC levels increased steadily when pasture on offer was high. When forage availability decreased in the mid-rainy season, WSC contents levelled off. It is possible that varying resting periods need to be utilised in order to optimise elephantgrass grazing management.

References
Passos, L. P., M. C. Vidigal, F. B. de Sousa, H. S. Barud, A. F. C. Paiva & A. R. Santos (2003). Comparative efficacy of autoclave-based extraction of soluble carbohydrates in various forage grasses. *In: World Conference on Animal Production, 9. Proceedings.* Porto Alegre, World Association of Animal Production, (CD-ROM).
Trevaskis, L. M., W. J. Fulkerson & K. S. Nandra (2004). Effect of time of feeding carbohydrate supplements and pasture on production of dairy cows. *Livestock Production Science*, 85, 275-285.

The role of grass tussocks in maintaining soil condition in north east Australia

B.K. Northup[1] and J.R. Brown[2]

[1]USDA-ARS Grazinglands Research Laboratory, El Reno, OK 73036, USA, Email: bnorthup@grl.ars.usda.gov,
[2]USDA-NRCS Jornada Experimental Range, Las Cruces, NM 88003, USA

Keywords: carbon, nitrogen, land degradation, state and transition

Introduction Soils of the grazing lands of north eastern Australia are inherently nutrient-poor. Heterogeneously distributed plants are important to the conservation of the limited amounts of nutrients, through storage in plant tissues or in soil sinks close to plants (Ludwig *et al.*, 1997). Loss of perennial vegetation through disturbance reduces conservation of these resources, to the detriment of feedback mechanisms, and ultimately causes loss of soil condition. Large areas of north east Australia have been degraded, or threatened by degradation, through combinations of variability in precipitation and heavy grazing (Gardener *et al.*, 1990). This study examined the inter-related responses of plants, soil microbes and soil nutrients to management-related disturbance.

Methods During the dry and wet seasons of 1998, replicate (n=4) soil samples from the upper 15 cm of the profile were collected in two sets of paddocks (n=2) in different condition (intact State 1 and degraded State 2), following 15 years (1983-1998) of grazing management [application of no and heavy (75% use of herbage produced each year) grazing, respectively]. Samples were collected at different locations (± 30 cm upslope-downslope) from the centre of tussocks of the two dominant perennial grass species (*Bothriochloa ewartiana* and *Chrysopogon fallax*). Total soil C and N were determined colorimetrically, and total microbial C was estimated from N values following ninhydrin fumigation of samples. A second set of samples was collected along similar transects to a depth of 50 cm, sectioned into 10 cm depth increments, roots were separated from soil, and surface area was defined by root scanner.

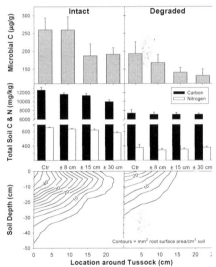

Results Microbial C, total soil C and N concentrations were present at higher levels in close proximity to grass tussocks on intact (State 1) condition paddocks (Figure 1). Concentrations on the degraded (State 2) paddocks were at or below the more distant locations of the intact paddocks. Microbial C was highest within ± 8 cm of centres of tussocks on both paddocks, compared with the remaining locations. Soil C and N declined with distance from tussock centres on the intact paddocks, while concentrations on the degraded paddocks were similar across locations. Root surface area was not widely distributed in either paddock, with the majority located within 20 cm of tussocks and at depths above 40 cm. Root surface areas in degraded paddocks were not recorded below 30 cm, nor in large amounts > 15 cm from tussock centres.

Conclusions The enrichment of soil by tussock grasses was highly localised, and highlights the inter-relatedness and tight coupling that exists between plants, nutrient pools, and microbial activity in north eastern Australia. Responses also underscore the importance of the biological component to

Figure 1 Microbial C, total C and N, (±1 s.e.) and root surface area distributions around grass tussocks

landscape condition. Disturbance of the herbaceous community had carry-over effects on components related to nutrient cycling and pools. Also, differences between the intact and degraded paddocks were not large [2150 (±520) mg/kg C, 65 (±10) mg/kg N, and 66 (±11) µg/g microbial C], indicating that reserves available to resist or recover from disturbance were limited. The differences noted here represent the effects of management over relatively short periods (15 years), and underscore the importance of balancing grazing pressure against productive capacity of the landscape.

References

Gardener C. J., J. G. McIvor & J. Williams (1990). Dry tropical rangelands: solving one problem and creating another. *Proceedings, Ecological Society of Australia*, 16, 279-286.

Ludwig, J., D. Tongway, D. Freudenberger, J. Noble, & K. Hodgkinson (1997). *Landscape Ecology, Function and Management: Principles From Australia's Rangelands*. CSIRO Publishing, Collingwood, Australia.

Effect of a grazing intensity gradient on primary production and soil nitrogen mineralisation in a humid grassland of western France

N. Rossignol, A. Bonis and J-B. Bouzillé
*UMR CNRS 6553 Ecobio, Campus de Beaulieu, Avenue du général Leclerc, 35042 Rennes Cedex, France,
Email: nicolas.rossignol@univ-rennes1.fr*

Keywords: primary production, grazing intensity, nitrogen mineralisation, humid grassland

Introduction Large herbivores have a major influence on the structure and the functions of humid grasslands and especially on primary production. Earlier work on the study site showed that grazing intensity was spatially varied and created a diversity of vegetation patches in the grassland (Loucougaray, 2003). The first objective of this study was to determine whether the variation in grazing intensity led to variation of primary production within the three plant communities located at three topographic levels in the grassland. The second objective was to determine whether a relationship linked primary production variation and net soil nitrogen (N) mineralisation.

Materials and methods For each vegetation patch, above ground net primary production (ANPP) was measured as the sum of positive biomass increments during four consecutive 15 day periods (May to June 2002) using temporary enclosures. Rates of net N mineralisation were measured in the first 15cm of the topsoil for each vegetation patch using the *in situ* buried bags method. Net N mineralisation rates were calculated as the difference between soil mineral N content at the beginning and end of an incubation period and were summed for 6 successive incubation periods (April to September 2002).

Results ANPP varied significantly between the plant communities and between the vegetation patches within the communities (nested ANOVA, $p < 0.001$). ANPP significantly decreased as the grazing intensity increased within each plant community (Figure 1). The decrease of ANPP along the gazing intensity gradient was stronger within meso-hygrophilous communities than within mesophilous and hygrophilous communities. Net N mineralisation rates varied significantly between vegetation patches. There was a linear negative relationship between ANPP and net N mineralisation rates (Figure 2, $R^2 = 43.7$ $p < 0.01$).

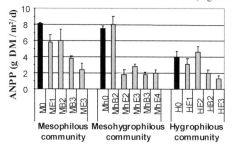
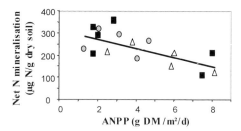

Figure 1 Effect of increasing grazing intensity on the ANPP of three plant communities. Values for ungrazed vegetation are reported in black.

Figure 2 Relationship between ANPP and net N mineralisation. Mesophilous (triangles), meso-hygrophilous (squares) and hygrophilous (circles)

Conclusions Within grassland, the variation in grazing intensity led to variation of above ground net primary production between and within plant communities. The results of this study show that there is a linear negative relationship between ANPP and net N mineralisation variation at the grassland scale. We suggest that the variations of grazing intensity influence soil carbon availability for soil microorganisms through variation of ANPP (Holland & Detling, 1990). The decrease in soil carbon availability leads to a decrease in the biomass of soil microorganisms. Such a decrease in microbial biomass could lead to a decrease in nutrient immobilisation by microbes and a subsequent increase in net N mineralisation.

References

Holland, E. A. & J. K. Detling (1990). Plant response to herbivory and belowground nitrogen cycling. *Ecology,* 71, 1040-1049.
Loucougaray, G. (2003). Régimes de pâturage et hétérogénéité de la structure et du fonctionnement de la végétation prairiale (Marais Poitevin). PhD Thesis, Université de Rennes 1, France.

Diet effects on dairy manure nitrogen excretion and cycling

J.M. Powell[1] and T.H. Misselbrook[2]
[1]U.S. Department of Agriculture, Agricultural Research Service, Dairy Forage Research Center, 1925 Linden Dr. West, Madison, Wisconsin 53706, USA, Email: jmpowel2@wisc.edu, [2]Institute of Grassland and Environmental Research, North Wyke, Okehampton, Devon, EX20 2SB, UK

Keywords: dairy diet, crude protein, manure, N cycling

Introduction The amount and forms of (nitrogen) N excreted by ruminant livestock and post excretion manure N cycling are highly influenced by what is fed. For example, the relative amount of urinary N, faecal endogenous N of microbial and gut origin, and faecal undigested feed N is affected by how much dietary fibre and secondary compounds (e.g., tannins, polyphneolics) are consumed. Each manure N component has a different propensity for loss; for example via ammonia (NH_3) volatilisation (Misselbrook *et al.*, 2004), leaching, and cycles differently in the soil-plant continuum (Powell, 2003). We evaluated dairy diet effects on amount and forms of manure N excreted and post excretion cycling of manure N from different diets after application to soil.

Materials and methods Holstein cows were fed different levels of crude protein (CP), fibre, corn silage, alfalfa silage, alfalfa haylage, and tannin-containing forages [alfalfa, birdsfoot trefoil low tannin (BF-T-Low) and birdsfoot trefoil high tannin (BF-T-High)] for the principal purpose of evaluating diet impacts on milk production and composition. At the end of each lactation trial, 3-4 cows per diet were fitted with indwelling catheters and urine and faeces were collected separately at *ca* 8 h intervals between two daily milkings for a total of 60-96 h. Diet impacts on NH_3 volatilisation (Table 1) were evaluated by recombining urine and faeces in the ratios they were excreted, applying their fresh or stored slurries to the surface of soils contained in laboratory chambers, and measuring NH_3 trapped in acidic solution over 48 hours. Diet impacts on soil N cycles (Figure 1) were evaluated by applying faeces from different CP-fibre-silage-haylage diets to potted soil at a rate of 350 kg N/ha and oats and sudangrass were grown in succession for 135 d.

Table 1 Predicted maximum cumulative NH_3 loss from fresh and stored slurries from different dairy diets on silt loam soil

Trial type	Trial components	Slurry type	
		Fresh	Stored
		% applied N volatilised	
CP level	13.6%	31b	12b
	19.4%	68a	29a
Forage tannin type	Alfalfa	31a	30a
	BF-T-Low	33a	23b
	BF-T-High	25b	19b

\# within each trial, values with different letters are significantly different (*P*<0.05)

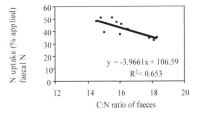

Figure 1 Relationship between C:N ratio of faeces and net plant uptake in silt clay loam (each data point = one diet)

$$y = -3.9661x + 106.59$$
$$R^2 = 0.653$$

Results Most observed diets had no impact on milk production but affected the amount, relative N partitioning, and composition of urine and faeces (data not shown). Fresh and stored slurry from the low CP diet had less than one-half the NH_3 loss than slurries from the high CP diet (Table 1). Fresh slurry derived from BF-T-High had less NH_3 loss than slurry from alfalfa or BF-T-Low diets. Stored slurry from BF-T-High and -Low diets had less NH_3 loss than slurry derived from alfalfa. Diets had significant impact on C:N ratios of faeces. As faecal C:N ratio increased, net N mineralisation and N uptake by oats and sudangrass decreased (Figure 1).

Conclusions The amount of CP and type of forage fed to lactating dairy cows had significant impacts on NH_3 loss. Diets also impacted faecal C:N ratios which apparently affected faecal N mineralisation in soil and subsequent plant N uptake. Diets could be formulated to meet nutritional requirements of high producing dairy cows and produce excreta less susceptible to environmental loss.

References

Misselbrook, T. H., J. M. Powell, G. A. Broderick & J. H. Grabber (2004). Reducing ammonia losses from dairy manure application to soil through dietary manipulation. Abstract 6220. In: *Agronomy Abstracts (CD)*. American Society of Agronomy. Madison, WI.

Powell, J. M. (2003). How dairy diet affects manure nitrogen excretion and cycling in soils. In: *Proc. of the 2003 Wisconsin Fertiliser, Aglime & Pest Management Conference.* 21-23 January 2003, Madison, WI, 308-314.

How will removal of the non-organic feed derogation affect nutrient budgets of organic livestock farms in Wales?

H. McCalman and S.P. Cuttle
Institute of Grassland and Environmental Research, Plas Gogerddan, Aberystwyth, Ceredigion, SY23 3EB, UK,
Email: heather mccalman@bbsrc.ac.uk

Keywords: organic farming, P and K nutrient budget, organic livestock, extension, technology transfer

Introduction Agri-environment schemes encourage organic farming in Wales. The National Assembly Government second organic action plan has a target of 10-15% land in organic production by 2010. Although forage based, many organic livestock farmers buy in concentrate feeds, which are important sources of nutrients to the farm. The current derogation allowing organic producers up to 10% approved non-organic feed ends in 2005. Many organic farmers are beginning to make management changes to comply with this. The aims of this study were to construct simple nutrient budgets on different organic livestock farms, investigate any planned system changes, look at impacts on nutrient budgets and create awareness amongst farmers of the value of nutrient budgeting, in addition to soil testing, for planning sustainable organic farming systems.

Materials and methods Data were collected from six commercial livestock farms in Wales (Farms 1 - 6). Simple farm-gate P and K budgets were determined for each farm and included data from the organic dairy research farm at IGER, Trawsgoed, that compares a purchased concentrates system (PF) with one that is largely self-sufficient in feed (SS). Farmers were asked to describe any management changes that they are likely to make to meet the feed regulations. The budgets were used to estimate the proportion of the nutrient input originating from purchased feed and how the removal of purchased feeds would affect the nutrient budget.

Results Under their current managements, all but two farms had a small P deficit (Table 1). Farm 3, the only one to apply P fertiliser, had a P surplus when the input from feed was excluded. All farms except SS had a K surplus under their current management. Excluding imported feed from the K budgets resulted in K deficits for four farms. However, Farms 3 and 5, which had applied K fertiliser or imported large amounts of straw, still had significant surpluses. Most farmers expected to reduce the amount of feed they purchased from outside the farm. Table 1 indicates that this would moderately increase deficits on some farms, although commercially the changes would be less marked as farms are unlikely to stop importing feed altogether and a reduction in nutrient inputs would generally be offset by a reduction in outputs. Other farms might sell more organic feed, which would also increase P and K offtakes and require increased fertiliser inputs within the organic sector.

Table 1 P and K surplus or deficit (input - output) from organic livestock farm farm-gate budgets and the impact of excluding purchased feeds from the budget (as kg/ha per year averaged over the whole farm area)

Farm number	1	2	3	4	5	6	SS	PF
Farm type[1]	BS	D/BS	BS	BS	D	D	D	D
	P surplus or deficit (kg/ha per year)							
Input - output	-1.9	-1.7	6.8	-0.7	-1.5	-5.1	-2.0	3.9
- excluding feed	-2.3	-4.1	4.8	-2.0	-5.7	-5.5	-3.4	-5.1
	K surplus or deficit (kg/ha per year)							
Input - output	5.3	0.1	18.6	9.8	24.8	0.7	-0.6	7.8
- excluding feed	4.2	-3.8	12.9	4.7	19.3	-6.9	-2.2	-3.5

[1] BS = beef/sheep; D = dairy

Conclusions The requirement to feed 100% organic feed from 2005 will result in a range of responses from organic farmers from 'no change' to radically modified systems. Many farmers indicated that they will reduce the amount of purchased feed. This would reduce inputs of P and K, but effects on nutrient budgets are likely to be relatively small except on those farms currently importing close to the maximum permitted amount of feed. This preliminary study has stimulated interest amongst the participating farmers and will be explored with them at discussion group meetings to highlight comparisons of farm type, year to year variation and the value of this type of benchmarking in technology transfer.

Acknowledgement The work was funded by Grassland Technology Transfer and Farming Connect projects.

Section 2

Chemical controls over soil quality and nutrient turnover

The effect of a reduction in phosphate application on soil phosphate pools

C. van der Salm[1], J. van Middelkoop[2] and P.A.I. Ehlert[1]

[1]Alterra, Wageningen University and Research Centre, P.O. Box 47, 6700 AA Wageningen, The Netherlands, Email: caroline.vandersalm@wur.nl, [2]Animal Science Group, P.O. Box 2176, 8203 AD Lelystad, The Netherlands

Keywords: phosphate pools, environmental P losses, dairy farms

Introduction Excessive use of manure and fertilisers in western Europe has led to high phosphorus (P) contents in many agricultural soils leading to environmental P losses by overland flow, subsurface drainage and leaching to groundwater. To stop phosphate build up in the soil and leaching to surface and ground waters, the Dutch government is gradually reducing allowable phosphate application on grassland from 130 kg/ha per year in 2005 to 90 kg/ha per year in 2015. This will lead to a reduction of the phosphate surplus from 40 in 2005 to 0 kg/ha per year. To investigate the impact of reductions in application rates on soil phosphate, leaching and grass production, a field experiment was started in 1997 on four dairy farms on two sandy soils, a peat and a clay soil.

Materials and methods At each dairy farm, six plots were established receiving a yearly N surplus of 180 and 300 kg/ha per year and P surpluses of 0, 20 and 40 kg P_2O_5/ha per year. The grass was alternately harvested and grazed. Fertiliser and manure inputs and grass production were measured regularly. Changes in P pools were monitored yearly in the topsoil (0-5, 5-10, 10-20 and 20-30 cm). Pw was measured in a 1:60 extraction with water, and P-Al by extraction with a 0.1 N ammonium lactate/0.4 N acetic acid solution. Changes in Pw and P-Al pools were assessed by multiple linear regression.

Results All sites had relatively high phosphate contents at the beginning. In the peat and clay soils, phosphate contents declined with depth, whereas contents were high throughout the topsoil (0-30 cm) of the sandy soils (Table 1). Phosphate losses from by leaching and drainage ranged from 1- 11 kg P_2O_5/ha per year total –P and 1-4 kg P_2O_5/ha per year MRP-P. These data indicate that a decline in P pools may be expected at a P surplus of 0 kg P_2O_5 ha per year and a rise in P pools will occur at P surpluses of 20 and 40 kg P_2O_5 ha per year. Changes in P pools in the different layers will be governed by the P surplus and the P losses to greater depth.

Table 1 Phosphate pools[1] in the topsoil of four farms in 1997

Site	Texture	Pw	P-Al	DPS[2]	P losses by leaching/drainage[3]	
					MRP-P	Total-P
		mg P_2O_5/ 1	mg P_2O_5 /100 g	(-)	kg P_2O_5/ ha per year	
Cr	Sand	40-17	40-28	0.44-0.39	3	6
Hn	Sand	39-24	50-38	0.59-0.58	4	11
Wb	Clay	56-8	58-12	0.46-0.18	0.7	1.3
Zg	Peat	31-4	42-5	0.36-0.16	1.2	3

[1] Contents in the 0-5 cm layer and the 20-30 cm layer resp. [2] $P_{ox}/0.5(Al+Fe)_{ox}$ [3] Average values 1997-2001

Multiple linear regression analysis confirmed this hypothesis. The change in Pw and P-Al was described by: $\Delta P = C_{soil} + \alpha\ P_{initial} + \beta\ P_{surplus}$ in the upper soil layer (0-5 cm) and by $\Delta P = C_{soil} + \alpha\ P_{initial} + \beta\ Pw_{x-1}$ in the deeper layers where ΔP represents the yearly change in P pool, C_{soil} a constant depending on the soil type, $P_{initial}$ the P pool in the previous year and Pw_{x-1} is the Pw of the overlying horizon. Pw_{x-1} is included as a measure for the P leaching from the overlying horizon (Schoumans & Groenendijk, 2000). The explained variance was 54.3 and 54.1 for ΔPw and 46.8 and 45.9 for ΔP-Al in topsoil and deeper layers, respectively. ΔP strongly increased with a decline in $P_{initial}$ (α: -0.9/-1.0 and -0.6/-0.7 for Pw and P-Al in top- and subsoil, respectively). ΔP increases with an increase in $P_{surplus}$ in the topsoil (β: 0.16 for Pw and P-Al).

Conclusions Regression equation indicated that Pw and P-Al values (0-5 cm) can be kept at a value of 35 at a P_2O_5 surplus of 0 kg/ha per year in sand and peat soils. To retain higher P levels (40), a P_2O_5 surplus of 20 kg/ha per year is required. In the clay soil, Pw and P-Al pools can be kept 20 units higher at the same P_2O_5 surplus. In the deeper soil layers, changes in P pools were independent of the surplus and remained stable over 7 years. Despite declines in Pw and P-Al in the topsoil at the lowest surpluses, grass production remained stable at 11 ton dry matter/ha per year. However, P content in the grass showed a slight decline on the sand and peat soils.

References

Schoumans, O. F. & P. Groenendijk (2000). Modelling soil phosphorus levels and phosphorus leaching from agricultural land in the Netherlands. *Journal of Environmental Quality*, 29, 111-116

Changes in nutrient turnover and supply during the reversion of arable land to acid grassland/*Calluna* heathland

A. Bhogal[1], B.J. Chambers[1], R. Pywell[2] and K. Walker[2]
[1]ADAS Gleadthorpe Research Centre, Meden Vale, Mansfield, Notts. NG20 9PF, UK, [2]Centre for Ecology and Hydrology, Monks Wood, Abbots Ripton, Huntingdon, Cambs., PE17 2LS, UK, Email: anne.bhogal@adas.co.uk

Keywords: heathland re-creation, soil pH, nutrient supply, sulphur

Introduction Lowland heath is of high conservation value because of the specialised and rare assemblages of plants and animals that it supports. Combinations of agricultural and urban development, and lack of appropriate management have resulted in large-scale loss and fragmentation of this habitat throughout the UK. Current UK conservation policies seek to re-create 6,000 ha of this habitat on land previously in agricultural and forestry production. Previous research indicated that high soil pH and fertility, together with a lack of propagules of heathland species, made it difficult to achieve this objective. The aim was to evaluate techniques to establish grass-*Calluna* heathland on ex-arable land in the Brecklands Environmentally Sensitive Area (ESA) in Norfolk, in particular, to assess the need for soil acidification to reduce competition and aid the establishment of *Calluna*.

Materials and methods Between 1994 and 1995 a number of different restoration treatments were applied to two ex-arable sites with contrasting soil pH and fertility levels in the Brecklands ESA: Euston (soil pH 5.7; ADAS Phosphorus Index 3) and Honington (soil pH 8.0; ADAS Phosphorus Index 4). These included elemental sulphur (S) additions (at two rates) to acidify the soil in combination with contrasting establishment techniques of a sown seed mixture (*Calluna* litter and acidic grassland species). Topsoil (0-15 cm) samples were taken periodically between 1994 and 2003 and analysed for pH, extractable phosphorus (P), potassium (K) and magnesium (Mg), cation exchange capacity (CEC) and potentially mineralisable nitrogen (PMN, 2003 samples only). Additional soil samples were taken to 90 cm depth and analysed for soil mineral N (ammonium-N plus nitrate-N, 1994-1998). Porous ceramic samplers were also installed on selected treatments at 90cm depth to measure nitrate concentrations in the drainage water during winters 1994/95 – 1998/99.

Results Elemental S was very effective in reducing soil pH in the first 12 months after application (Figure 1a). Thereafter, pH levels gradually increased to reach an equilibrium of *ca* 4.4 and 4.2 at Euston (3 & 6 t/ha S) and *ca* 5.0 and 4.1 at Honington (9 & 18 t/ha S) during the period 1999-2003. The soil pH of the untreated control treatments remained relatively constant throughout at *ca* 5.6 at Euston (range 5.2-6.2) and *ca* 7.8 at Honington (range 7.4-8.0). *Calluna* established only on the acidified S treatments (Figure 1b).

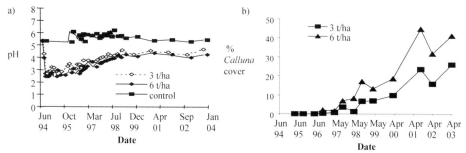

Figure 1 Effects of restoration treatment on a) topsoil pH and b) the mean % cover of *Calluna vulgaris* (Euston)

Soil acidification resulting from elemental S additions, led to long-term increases in soil extractable P at both sites (by 20-76 mg/l; $P<0.001$), but PMN and extractable Mg were reduced (by 22-57 mg/kg and 7-18 mg/l, respectively; $P<0.001$). In addition, extractable K was reduced by 38-50 mg/l at Honington ($P<0.001$). The higher rate of S addition at each site increased SMN in each measurement year, compared with the control, which was largely due to a larger ammonium-N pool, suggesting that nitrification had been inhibited at low soil pH. Reverting arable land to acid grassland/*Calluna* heath was very effective in reducing nitrate leaching losses; losses in drainage water were generally <5 kg/ha N and mean nitrate-N concentrations below 5 mg/l.

Conclusions Soil acidification was required for the recreation of heathland on ex-arable land in the Brecklands ESA. Elemental S was very effective in reducing soil pH and also changed other soil nutrient supply properties.

Study of dairy manure N cycling in soil-plant continuum using ^{15}N and other methods

J.M. Powell[1], P.R. Cusick and K.A. Kelling
U.S. Department of Agriculture, Agricultural Research Service, Dairy Forage Research Center, 1925 Linden Dr.
West, Madison, WI 53706, USA, Soil Science Department, 1525 Observatory Dr., University of Wisconsin,
Madison WI 53706, USA, Email: jmpowel2@wisc.edu

Keywords: dairy manure, ^{15}N, N cycling

Introduction Ruminant livestock manure impacts on N cycling in the soil-plant continuum. Most studies of manure N cycling are short-term and rely on indirect methods, i.e. apparent N recovery, fertiliser N equivalents or incorporate ^{15}N into ammonium-N fractions. Direct and perhaps more precise measurements may be achieved by long-term studies using ^{15}N incorporated into all manure N components. This paper summarises results of a 6-year trial to compare indirect and direct measures of manure N uptake by corn for 3 years after application.

Materials and methods ^{15}N enriched dairy manure, obtained by feeding ^{15}N enriched forage (Powell et al., 2005) was applied to corn (Zea mays) plots on a Plano silt loam (Typic Argiudolls) at rates equivalent to 210 kg N/ha every 1, 2 or 3 years in south central Wisconsin (45° 05' N, 89° 31' W) during 1998-2003 (Cusick, 2004). There were also non-manure control and fertiliser N (0, 45, 90, 135, 179 and 224 kg N/ha as NH_4NO_3) plots. Apparent manure N recovery was determined as differences (DIFF) in corn N uptake in manure-amended and control plots, divided by applied manure N. Fertiliser N equivalent (FE) was interpolated from applied fertiliser N (x) and total corn N uptake (y) regressions. Manure N availability was determined by dividing FE by manure N applied.

Results 1st, 2nd and 3rd year estimates of manure N uptake by corn (Table 1) were 25, 6 and 2% in plots amended with manure ^{15}N; 25, 6 and 0% by DIFF; and 55, 24 and 25% by FE. Although manure ^{15}N and DIFF provided similar estimates, the variability associated with DIFF was much greater, especially for years 2 and 3. Estimate variabilities were greatest with FE. Low manure ^{15}N uptake by corn was due to high background fertility of the plots, which retained 46% of applied manure ^{15}N in the upper 30 cm of soil 3 years after application. Drawbacks of using ^{15}N are labour, cost and advance planning to develop labelled diets. Short-term (<2 years) studies can use manure ^{15}N by feeding urea to cows (Powell et al., 2005). "Urea manure" contains labelled urine- and faecal microbial-N (ca 50 and 40% of excreted manure N, respectively). The present manure also contains labelled faecal undigested feed N (ca 10% of excreted manure N). Both types gave similar 1st and 2nd year estimates of corn ^{15}N uptake and ^{15}N in soil total and inorganic N (Powell et al., 2005).

Table 1 Estimates of dairy manure N uptake by corn using ^{15}N manure, difference method (DIFF) and fertiliser equivalent (FE). Values are means (standard error) of four replicates per year (adapted from Cusick, 2004)

Manure application	Corn N uptake 1st year (% of applied manure N)		
	^{15}N	DIFF	FE
Single (5 yr)	19 (3.8)	22 (6.9)	52 (13.0)
2 consec. (1 yr)	18 (3.6)	12 (4.0)	17 (7.8)
3 consec. (1 yr)	26 (8.3)	13 (4.8)	57 (14.0)
4 consec. (1 yr)	37 (8.3)	50 (11.3)	75 (16.9)
5 consec. (1 yr)	30 (3.9)	28 (10.8)	102 (1.9)
6 consec. (1 yr)	19 (3.4)	22 (8.4)	28 (16.1)
	2nd year (% of applied manure N)		
Single (5 yr)	6 (1.1)	6 (9.7)	24 (9.8)
	3rd year (% of applied manure N)		
Single (4 yr)	2 (0.4)	0 (7.5)	25 (8.5)

Conclusions Indirect estimates of manure N uptake by plants (i.e. DIFF and FE) are less precise than direct measurements using manure ^{15}N. Long-term manure-soil-plant-N cycling studies require manure ^{15}N derived by feeding ^{15}N-labelled feed; short-term studies may use manure ^{15}N derived from feeding ^{15}N-labelled urea.

References

Cusick, P. R. (2004). Residual dairy manure nitrogen availability and mineralization of whole and individual manure nitrogen components. MS Thesis, Soil Science Department, University of Wisconsin-Madison.
Powell, J. M., K. A. Kelling, G. R. Muñoz & P. R. Cusick (2005). Evaluation of dairy manure ^{15}N enrichment methods on short-term crop and soil nitrogen budgets. Agronomy Journal, 97, 333-337.

Nitrogen leaching from cattle, sheep and deer grazed pastures in New Zealand

K. Betteridge[1], S.F. Ledgard[2], C.J. Hoogendoorn[1], M.G. Lambert[1], Z.A. Park[1], D.A. Costall[1] and P.W. Theobald[1]
[1]AgResearch Grasslands, Private Bag 11008, Palmerston North, NewZealand, Email: keith.betteridge@agresearch.co.nz and [2]AgResearch Ruakura, Private Bag, Hamilton, New Zealand

Keywords: nitrate leaching, deer, sheep, cattle, grazing

Introduction The impacts of intensified grazing in New Zealand are being reflected in declining quality of groundwater, streams and lake water. Manipulation of ratios of grazing animal species may be one way farmers can reduce nitrogen (N) emissions to ground water. The present research quantifies nitrate and ammonium leaching losses from rotationally grazed sheep, cattle and deer pastures in a common environment.

Materials and methods On a highly porous pumice soil, triplicate 0.4 to 0.6 ha plots were fenced to contain 5 to 17 month-old female cattle, sheep or deer. Grazing started in October 2003 and pasture mass was measured before and after 2-day grazing events to determine pasture utilisation. The pre-grazed pasture was analysed for N content to allow determination of N intake by grazing animals. Animal grazing-days were recorded and expressed as sheep stock unit equivalents (SU). Within plots, 40 randomly located ceramic cup samplers, set 60 cm below the soil surface, were used to sample drainage water, between March and October 2004. Nitrate and ammonium concentrations in drainage water were assayed for each of 7 sampling times. Drainage volume was estimated by water balance estimations.

Results During the 8 months from March 2004, pasture and N utilisation and SU grazing days were similar across treatments (Table 1). During this period, 700 mm of the 950 mm rainfall drained to groundwater. At similar levels of pasture dry matter and N utilisation, nitrate leaching losses over 8 months from young, non-lactating cattle grazed swards was more than twice that from swards grazed by either sheep or deer (Table 1). Of the N apparently consumed, 22% was leached as nitrate N, compared with 10 to 11 % for deer and sheep. Only 0.4 kg/ha of ammonium N was leached from each pasture. Nitrate losses from these sheep and dairy pastures on pumice soils, without fertiliser N, are high compared with losses from sheep or dairy grazed pastures without fertiliser N, on sedimentary or less-free draining volcanic ash soils, where drainage volumes were substantially less (Ruz-Jerez *et al.*, 1995; Ledgard *et al.*, 1996; Eckard *et al.*, 2004).

Table 1 Eight month mean pasture and herbage N utilisation, sheep stock unit equivalent SU grazing days and leached nitrate N

Treatment	Utilised DM (kg/ha)	Utilised N (kg/ha)	SU grazing Days	Nitrate-N (kg/ha)	Nitrate-N/ utilised N (%)
Cattle	4320	159	4106	35.0 a	21.6 a
Sheep	3960	160	4150	17.8 b	11.1 b
Deer	4930	167	3924	15.8 b	9.5 b
LSD$_{0.05}$	2133	89	1078	15.7	8.9

Conclusions Increasing the proportion of feed utilised by sheep and deer, and decreasing the proportion utilised by cattle will reduce nitrate leached to ground water.

References
Eckard, R. J., R. E. White, R. Edis, A. Smith & D. F. Chapman (2004). Nitrate leaching from temperate pastures grazed by dairy cows in south-east Australia. Australian Journal of Agricultural Research, 55, 911-920.
Ledgard, S. F., J. W. Penno & M. S. Sprosen (1999). Nitrogen inputs and losses from clover/grass pastures grazed by dairy cows. Journal Agricultural Science (Camb.), 132, 215-225.
Ruz-Jerez, B. E., R. E. White & P. R. Ball (1995). A comparison of nitrate leaching under clover-based pastures and nitrogen fertilised grass grazed by sheep. Journal of Agricultural Science (Camb.), 125, 361-360.

Effect of soil chemistry on microbial biodiversity and functionality in grassland and tilled soils

C. Carrigg[1], S. Kavanagh[1], D. Fay[2] and V. O' Flaherty[1]
[1]Microbial Ecology Laboratory, Environmental Change Institute, National University of Ireland, Galway, Ireland, Email: cora.carrigg@nuigalway.ie, [2]Teagasc, Johnstown Castle, Co. Wexford, Ireland

Keywords: soil microbiology, DGGE-PCR Denaturing Gradient Gel Electrophoresis, Polymerase Chain Reaction, ammonium oxidisers

Introduction Microorganisms are excellent indicators of soil health, because of their rapid response to environmental change. Traditional microbiology is ineffective for the study of soil, as <1% of microorganisms are currently culturable (Torsvik *et al.*, 1996). Nucleic acid based methods, however, allow rapid detection of organisms, or particular genes, directly from soil samples. This work investigated, using polymerase chain reaction (PCR)-based approaches, the relationship between key chemical properties and bacterial biodiversity in grassland and tilled soils, with particular emphasis on the abundance and diversity of ammonium oxidisers.

Materials and methods Thirty grassland soils, and 5 soils under barley (pH from 5.1-7.2), were characterised for a range of key chemical properties: those for one soil (844) were: lime, 10.1 t/ha; P 3.3 mg/l; K 300 mg/l, Mg 109 mg/l; pH 5.1; organic C 4.79%; and total N 0.59%. DNA was extracted with a rapid method developed at NUI, Galway, ensuring a cell lysis efficiency of >80%. Bacterial biomass was calculated based on DNA yield. Denaturing Gradient Gel Electrophoresis (DGGE)-PCR was carried out with 16S rRNA gene-specific primers for all bacteria (Muyzer *et al.*, 1993) and *amo*A gene primers for ammonium oxidisers (Holmes *et al.*, 1995). PCR amplicons were run on DGGE gels to generate banding patterns. Un-weighted Pair Group with Arithmetic Mean (UPGMA) analysis of DGGE profiles was carried out with *MEGA* 2.1 (Kumar *et al.*, 2001; Figure 1).

Results This study demonstrated that the microbial biomass and biodiversity in grassland and tilled soils can be related to basic soil properties (Figure 1), as could the functional capacity of soils with respect to ammonium-oxidation. This clear relationship between soil type, vegetation type, land use and microbial communities has significant implications for the study and understanding of nutrient cycling and organic matter decomposition.

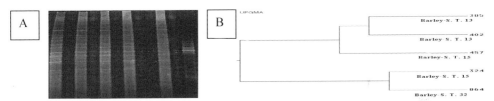

Figure 1 (A) DGGE gel from 5 soil samples under barley. The number and position of bands are indicative of bacterial species within the soil community, common bands are clearly visible. (B) UPGMA analysis of DGGE profiles. The microbial biodiversity of soils could be statistically related to properties such as pH.

Conclusions Microbes play many key roles in soil, but major knowledge gaps exist in our understanding of soil microbiology. Rapid, statistically verifiable, microbial analysis of soils is now feasible for large-scale applications and may be integrated into soil monitoring and research programmes.

References
Holmes, A., J., A. Costello, M. E. Lidstrom & J. Murrell (1995). Evidence that particulate methane monooxygenase and ammonia monooxygenase may be evolutionarily related. *FEMS Microbiology Letters,* 132, 203-208.
Kumar, S., K., K. Tamura, I. B. Jakobsen & M. Nei (2001). *MEGA2*: Molecular Evolutionary Genetics Analysis software. *Bioinformatics,* 17, 1244-1245.
Muyzer, G. E. C., de Waal & A. G. Uitterlinden (1993). Profiling of complex microbial populations by denaturing gradient gel electrophoresis analysis of polymerase chain reaction-amplified genes coding for 16S rRNA. *Applied & Environmental Microbiology,* 59, 695-700.
Torsvik, V., R. Sorheim & J. Goksoyr (1996). Total diversity in soil and sediment communities – a review. *Journal of Industrial Microbiology,* 17, 170-178.

Effect of different carbon and nitrogen inputs on soil chemical and biochemical properties in maize-based forage systems in Northern Italy

S. Monaco[1], D. Hatch[2], L. Dixon[2], C. Grignani[1], D. Sacco[1] and L. Zavattaro[1]

[1]Dipartimento di Agronomia, Selvicoltura e Gestione del Territorio -Università di Torino, Via L. da Vinci 44,10095 Grugliasco, Italy, Email: stefano.monaco@unito.it, [2]Institute of Grassland and Environmental Research, North Wyke Research Station, Okehampton, Devon, EX20 2SB, UK

Keywords: forage systems, nitrogen fertilisation, soil organic matter, soil mineralisable nitrogen

Introduction In agroecosystems, manure application and straw return affect carbon (C) and nitrogen (N) cycling and affect soil organic matter (SOM), nutrient supply and losses to the environment. We examined effects of different organic sources on crop production, N uptake and surplus and SOM in maize systems.

Materials and methods The experiment (established 1992), is located near Turin, northern Italy, on a deep calcareous sandy-loam soil in a completely randomised block design with 3 replicates. Treatments were maize: either silage (MS), or grain (MG); receiving N as liquid (LM) or solid (SM) manure, and/or urea at a nominal total input of 300 kg/ha (Table 1). N supply/uptake and crop production were measured each year. In 2003, soil organic C and total N contents were assessed and in 2004, the potential N availability SOM from mineralisation was assessed by anaerobic incubation (Waring & Bremner, 1964) and hot KCl extraction (Whitehead, 1981).

Table 1 Crop and fertiliser application to the treatments

Treatment	Crop	Liquid manure	Solid manure	Urea
		N kg/ha		
MS0	Maize silage	0	0	0
MS300	Maize silage	0	0	305
MG300	Maize grain	0	0	305
MS300LM	Maize silage	222	0	100
MS300SM	Maize silage	0	285	100

Table 2 Crop/*grain production, N uptake and balance: different letters indicate differences at P<0.05 (SNK test)

Treatment	Crop prod. DM t/ha		Uptake	Offtake	Balance
			N kg/ha		
MS0	16.4[a]		128.2[a]		-128.2
MS300	24.3[b]		248.7[b]		56.3
MG300	25.8[b]	13.3*	273.9[b]	185.6	119.4
MS300LM	25.3[b]		253.8[b]		67.8
MS300SM	24.9[b]		254.1[b]		131.3

Results Application of animal manures or crop residues had no effect on dry matter production or N uptake (Table 2). However, N balance was higher with SM from the extra N supplied, and with MG, from the crop residues on the soil surface. Treatments produced differences in SOM (Table 3). In particular, the larger C and N supply (from manure or maize straw returned to the soil) increased soil C and N contents. The two potential N mineralisation methods showed different results. Whilst amounts of N released during anaerobic incubation were clearly related to SOM accumulation, i.e. more mineralised N with higher soil C and N contents, N mineralisation through hot KCl extraction seemed to be more dependent on amounts of organic residues added: the higher values with higher rates of straw return in MG 300, or to higher rates of manure in MS300SM.

Table 3 Soil C, N and C/N ratio and potentially mineralisable N in the 0-30 cm layer: different letters indicate significant differences at P<0.05 (SNK test)

Treatment	C org	N tot	C/N	Anaerobic incubation	Hot KCl
	%	%		NO$_3$-N mg/kg	
MS0	0.95[a]	0.11[a]	8.57[a]	30.7	12.2
MS300	0.96[a]	0.12[a]	8.26[a]	23.6	18.2
MG300	1.12[ab]	0.13[a]	8.64[a]	42.5	23.5
MS300LM	1.15[b]	0.13[b]	8.64[a]	54.8	18.3
MS300SM	1.36[c]	0.15[c]	8.89[a]	66.7	21.0

Conclusions Treatments did not affect crop production or uptake, but caused differences in SOM. Manure application, and the return of straw increased soil C and N. Potential N mineralisation was related to soil C and N contents, but showed different ranking depending on the quality of accumulated SOM.

References

Waring, S. A. & J. M. Bremner (1964). Ammonium production in soil under waterlogged conditions as an index of nitrogen availability. *Nature*, 201, 951-952.

Whitehead, D. C. (1981). An improved chemical extraction method for predicting the supply of available soil nitrogen. *Journal of Science, Food and Agriculture*, 32, 359-365.

Seasonal changes in the ratio of microbial biomass P to total P in soils of grazed pastures

M. Kaneko, Y. Kurokawa, H. Tanaka and S. Suzuki
Tokyo University of Agriculture & Technology, 3-5-8, Saiwaicho, Fuchu, Tokyo, Japan,
Email: m2k2@cc.tuat.ac.jp

Keywords: microbial biomass P, seasonal change, P cycle, pasture

Introduction Phosphorus (P) utilisation efficiency in pasture soils is higher than in arable soils. Because there is a considerable amount of microbial biomass in the root mat layer, which is peculiar to permanent pasture, the microbial biomass P (MBP) contribution may be important in supplying soil P to pasture plants (Chen *et al.*, 2000; He *at al.*, 1997). In the present study, we investigated seasonal changes in MBP and other forms of P relative to total soil P in two pastures in which P uptake was estimated to be different.

Materials and methods This experiment was conducted at two pastures on Silic Andosols in Field Museum Tsukui (Kanagawa prefecture) of Tokyo University of Agriculture & Technology. A pasture which had been reseeded with tall fescue (TF, *Festuca arundinacea* Schreb.) two years prior to our study and a pasture of Japanese lawn grass (JL, *Zoysia japonica* Stend.) semi-natural pasture were used. N fertiliser was applied to the TF pasture on May 22. Six core soil samples (50 cc and 0-2.5cm in depth) were taken at each of three points in each pasture on April 16 (spring), July 17 (summer) and October 16 (autumn) in 2003. Fresh samples were sieved through a 2 mm screen and used to determine MBP and Olsen P (OP). MPB was determined with a $CHCl_3$ fumigation-0.5M $NaHCO_3$ extract method with pre-incubation. A portion of the sieved soil was air-dried and used to determine total P (TP, HNO_3-$HClO_4$ digestion method) and a modified Bray No. 2 method P (BRP). All P was determined colorimetrically by the ammonium molybdate method. MBP was calculated by $Kp = 0.4$.

Results and discussion Table 1 illustrates the seasonal changes in grazed pasture soil P content. The contents of TP ranged from 2.36 to 3.77 mgP/g dry soil. Differences in TP among seasons were not significant. Throughout the season, TP in the JL pasture was higher than that in the TF pasture ($p<0.01$). The proportions of MBP and BRP to TP in the JL pasture were also higher than those in the TF pasture ($p<0.01$), even though no fertiliser had been applied to the JL pasture for 10 years. The values of MPB percentage were similar to those presented by Chen *et al.*, (2000) BRP and OP contents were similar to those for other soils (Srivastava, 1992).

Conclusions It is possible that higher MPB in the JL pasture was caused by root mat development. MBP in the JL pasture reached a minimum in spring and maximum in autumn, while in the TF pasture, maximum MPB was observed in spring and minimum in summer. He *et al.* (1997) reported that MBP reached a minimum in summer: this is consistent with the TF pasture, and inconsistent with the JL pasture, although seasonal change in MPB was not significant. Srivastava (1992) suggested that water content is the most important factor in MBP change. In our study the relationship between MBP and soil water content was not clear. Significant seasonal change was found only in BRP of TF ($p<0.01$). Examination of changes in OP and BRP and the balance between MBP dynamics and P absorption by plants would enable us to utilise pasture soil P more efficiently.

Table 1 Seasonal change in pasture soil phosphorus

		Total P (mg P/g DS)	MBP (% TP)	OP (% TP)	BRP (% TP)	BRP-OP (% TP)	Soil water content g/g dry soil
Japanese lawngrass	Spring	3.31	2.2	2.9	4.9	2.0	0.57
	Summer	3.77	2.9	1.7	6.8	5.1	0.67
	Autumn	3.68	3.3	2.5	5.2	2.7	0.63
Tall fescue	Spring	2.36	1.8	1.5	1.6	0.1	0.60
	Summer	2.92	1.0	1.8	2.2	0.4	0.67
	Autumn	2.72	1.2	2.5	4.7	2.2	0.47

MBP: Soil Microbial biomass P, OP: Olsen P, BRP: Modified Bray No. 2 P

References
G. C. Chen, Z. L. He & C. Y. Huang (2000). Microbial biomass phosphorus and its significance in predicting phosphorus availability in red soil. *Communications in Soil Science and Plant Analysis,* 31, 655-667.
Z. L. He, J. Wu, A. G. O'Donnell & J. K. Syers (1997). Seasonal responses in microbial biomass carbon, phosphorus and sulphur in soil under pasture. *Biology and Fertility of Soils,* 24, 421-428.
S. C. Srivastava (1992). Microbial C, N and P in dry tropical soils: seasonal changes and influence of soil moisture. *Soil Biology and Biochemistry,* 24, 711-714.

Nitrogen mineralisation in situ and in controlled environment

F. Pálmason
The Agricultural Research Institute, Reykjavík, Iceland, Email: fridrik@rala.is

Keywords: net mineralisation *in situ*, nitrogen, andosol

Introduction Net mineralisation may be underestimated by *in situ* soil core methods for at least two reasons: (1) absence of plant N uptake in the soil cores, causing higher immobilisation than in intact soil. Schimel & Bennet (2004) thus concluded that plants compete effectively with microbes, as strongly indicated by cases, where net mineralisation was lower than plant uptake. (2) Gaseous losses of N_2O in closed cylinders can lead to underestimation especially during long incubation periods, Abril *et al.* (2001) and Vor & Brumme (2002).

Materials and methods Net mineralisation was estimated (a) in sieved soil, kept frozen from sampling until incubated at 15°C and water tension 10 k Pa and (b) by an *in situ* method, using 25 cm long steel cylinders inserted to 20 cm depth, closed above with a U tube to exclude leaching by rain. Incubation periods were 14 days. Net mineralisation exceeding uptake and losses from soil was estimated from soil samples of intact soil. Net mineralisation at 15°C and water tension at 10 k Pa was adjusted to field soil temperatures and water from the Arrhenius equation and with $Q_{10} = 2$, 4 and 8 and water content relative (%) to water content at 10 k Pa.

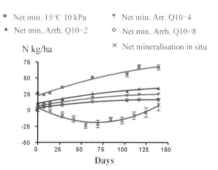

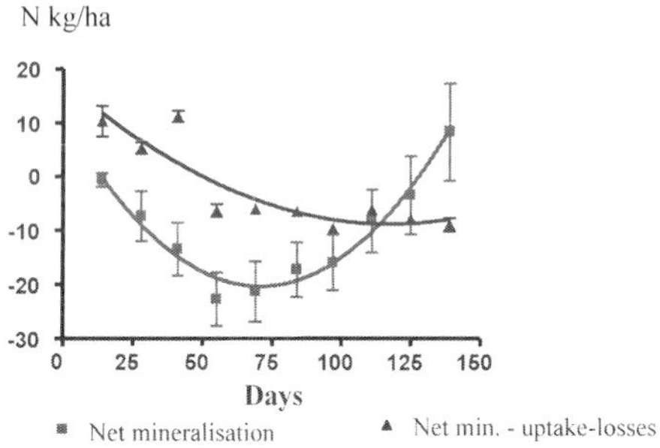

Figure 1 Net mineralisation (a) incubated in controlled environment, (b) adjusted to field temperature and water and (c) in soil cores *in situ*[1]

Figure 2 Net mineralisation in soil cores *in situ* and net mineralisation exceeding uptake and losses in intact soil[1]

[1]The graphs show regression lines for second order polynomials with standard error bars

Results Recalculation of net mineralisation at 15°C and 10 k Pa to temperatures and soil water tension in the field could only partly explain deviation from net mineralisation in soil cores (Figure 1). Remaining differences may be caused by denitrification in the enclosed cores and extra mineralisation from freezing and thawing of soil used in the laboratory incubation. The polynomial second order equations explain 82, 83, 93, 93 and 27% of the variation in net mineralisation in controlled environment, after adjusting to field condition with $Q_{10} = 2$, 4, 8 and *in situ*, respectively (Figure 1). Net N mineralisation exceeding uptake and losses in intact soil is significantly higher ($P = 0.0018$) than net mineralisation in *in situ* cores in the main growth period until day 84 (Figure 2) indicating net mineralisation was underestimated because of increased denitrification in the enclosed cores or by high immobilisation in the absence of N uptake by plants.

Conclusions Net mineralisation in an andosol soil was underestimated by an *in situ* method, most likely by increased denitrification and immobilisation of N in the enclosed soil cores as compared with intact soil. Short incubation periods and adequate aeration in the enclosed cores comparable to conditions in intact soil seem essential in order to minimise underestimation of net mineralisation by this method.

References
Abril, A., V. Caucas & E. H. Bucher (2001). Reliability of the *in situ* incubation methods used to assess nitrogen mineralisation: a microbiological perspective. *Applied Soil Ecology,* 17, 125-130.
Schimel, J. P. & J. Bennett (2004). Nitrogen mineralization challenges of a changing paradigm. *Ecology,* 85, 591- 602.
Vor, T. & R. Brumme (2002). N_2O losses result in underestimation of *in situ* determinations of net N mineralisation. *Soil Biology and Biochemistry,* 34, 541-544.

N-mineralisation and phosphorous: important elements in decision support for grassland systems

A.L. Nielsen[1] and C.C. Hoffmann[2]

[1]Natlan, Agro Business Park, 8830 Tjele, Denmark, Email: lisbeth.nielsen@agropark.dk, [2]National Environmental Research Institute, 8600 Silkeborg, Denmark

Keywords: organic soils, N-leaching, P-leaching, grazing intensity, cutting

Introduction Leaching of N and P from extensively managed grasslands on organic soils varies considerably. In environmentally sensitive areas it is important to diminish leaching by appropriate agricultural management. In Denmark low grazing intensity and management without fertilisation have been given a high priority. The type of soil has not been equally in focus, and it seems that the effect of cutting, compared with grazing, results in a higher removal of nutrients (e.g. Benke *et al.*, 1992) can be used more strategically. The objective of this case study was to combine data from management strategies with data from leaching studies on organic soils to elucidate the differences between type of management and type of soil for the potential leaching of N and P.

Materials and methods The study was carried out at two separate sites 2 km apart, referred to as 'West' and 'East'. At 'East' the effect of three management strategies on soil mineral N was examined in a block design with three replicates: a) continuous grazing with steers (compressed sward height (CSH): 6 cm), b) two cuts, c) two cuts with deep litter (20 t/ha) average 83 kg N of which 6 kg NH_4N, 17 kg P and 134 kg K. Leaching of N and P was recorded at high and low grazing intensity in 'West' and 'East'. Continuous high intensity grazing with steers and sheep in separate paddocks aimed at a CSH of 6 cm. At low grazing intensity the number of animals/ha was 50% (steers) or 65% (sheep) of the number at high intensity. The lowest level of the water table varied between 30 and 55 cm below soil surface in the years of the experiment (Hald *et al.*, 2003a).

Results Initial soil analyses for, respectively, 'West' and 'East' were $pH(CaCl_2)$ 5.6 and 4.7, total N (%) 1.8 and 2.7, soil organic matter (%) 49 and 66, C:N-ratio 15.6 and 13.9, N-mineralisation measured by incubation at $20°C$ 2.0 and 4.6 kg N/ha per day. Soil mineral N increased through the growing season on the grazed plots compared with the cutting treatments and soil mineral N in spring was lower in plots with cutting and deep litter compared with the other treatments (Table 1). Leaching of N and P was considerably higher in 'East' compared with 'West' (Table 2). There was no difference in N and P leaching between high and low grazing intensity.

Table 1 Soil mineral N (kg N/ha) with three management strategies ('East'), average of 1998-2000

	Soil sampling	Continuous grazing	Two cuts	Two cuts, deep litter	LSD*	Average of 20 grasslands**
Change through the growing season	0-20 cm	21.0	0.9	-1.0	20.8	-1
	20-40 cm	2.1	3.6	1.9	(7.2)	9
Mineral N in spring	0-20 cm	30.0	33.8	28.1	(9.6)	42
	20-40 cm	30.6	30.1	22.5	7.5	14

*LSD: Management, least significant difference (p<0.05), **Different management intensity (Hald *et al.* 2003b)

Conclusions This case study demonstrates that the level of N-mineralisation is important when making decisions about management. Where the level of N-mineralisation is high it is possible to remove N from soil when supplying the correct amounts of limiting nutrients. Where depletion of nutrients is required for the environment, cutting can be used for a number of years, but managements should be adjusted according to changes in soil conditions and the intentions for the area.

Table 2 Concentration of N and P in field drains*

		Mean	Std. Error	N**
Total N mg/l	'West'	2.6	0.13	69
	'East'	11.1	0.41	69
Total P mg/l	'West'	0.4	0.03	58
	'East'	1.8	0.14	58

*Data from two similar grasslands with 1.3 mg N/l and 0.13 mg P/l (Grant, R., pers. communication)
**N, number of samples

References

Benke, M., A. Kornher & F. Taube (1992). Nitrate leaching from cut and grazed swards influenced by nitrogen fertilization. Proceedings 14th General Meeting of the European Grassland Federation, 184-188.

Hald, A. B., C. C. Hoffmann & L. Nielsen (2003a). Ekstensiv afgræsning af ferske enge. DIAS Report 91, 191 pp.

Hald, A. B., A. L. Nielsen, K. Debosz & J. H. Badsberg (2003b). Restoration of degraded low-lying grasslands: indicators of then environmental potential of botanical nature quality. Ecological Engineering, 21, 1-20.

Implications for N transformations in acidic soils of replacing annual-based legume pastures with lucerne-based pasture in dryland farming systems of southern Australia

I.R.P. Fillery

CSIRO Plant Industry, PO Private Bag 5, Wembley WA 6913 Western Australia, Austrtalia, Email: Ian.Fillery@csiro.au

Keywords: N fixation, net soil N mineralisation, lucerne, alfalfa, subterranean clover

Introduction   The supply and demand for nitrogen (N) in annual-based pasture-crop rotations in southern Australia is often poorly synchronised, leading to large losses of inorganic N (Fillery, 2001). Perennial pasture species, particularly lucerne, are being recommended to minimise dryland salinity. The implications for N cycling of using lucerne in place of annual legumes on acidic sandy soils that are widespread have not been widely studied. Lucerne is less tolerant of acidity and could fix less N than annual legumes. Lucerne root residues mineralise at slower rates than annual pasture residues with lower N release to subsequent wheat crops (Bolger *et al.*, 2003). The aims of the work were to compare N cycling under lucerne and the traditional annual legume-based pasture system.

Materials and methods   Lucerne (*Medicago sativa*), and subterranean clover (*Trifolium subterraneum*) were sown into separate 60 m long by 20 m wide fenced plots that enabled grazing by sheep. Treatments were replicated four times. Dry matter production between grazing events, N content in pasture components (legume and non-legumes), ^{15}N natural abundance in legumes and non-legume reference plants and biomass were determined. Soil was sampled in layers to a depth of 1.6 m periodically during the growing seasons studied, and samples analysed for ammonium and nitrate (NO_3^{-1}). Net N mineralisation during the growing season of pasture phases and a subsequent wheat crop was determined using an *in situ* incubation technique. Rainfall and other climatic variables were also measured.

Results   The amounts of N fixed by legume shoots over an 18-month period that covered two growing seasons (April 1999 to October 2000) included 215 kg N/ha N in subterranean clover-based pasture and 157 kg N/ha for lucerne-based pastures, a statistically significant difference. Uptake of soil-derived N in subterranean pastures (197 kg N/ha) was largely undertaken by non-leguminous species (125 kg N/ha) whereas lucerne used 69 kg N/ha and non-leguminous species in lucerne-based pastures used 95 kg N/ha soil N. Lucerne was an important sink for soil-derived N in the summer-autumn of 2000 when it accumulated 31-35 kg /ha of soil derived N in shoot. The effect of this uptake on soil NO_3^{-1} levels ahead of winter rainfall is shown in Figure 1. Soil profiles under senesced subterranean clover contained 94 kg N/ha of NO_3^{-1} to 1.6 m compared with 45 kg N/ha under lucerne. Root uptake of soil-derived N was not determined, thus legume and non-legume uptake of mineral N are underestimated. Soil NO_3^{-1} under subterranean clover decreased through the winter growing season with the growth of non-leguminous species but plant recovery of NO_3^{-1} would have been lower had leaching of NO_3^{-1} occurred. Net N mineralisation in soil for the period April 1999 to October 2000 amounted to 178 kg N/ha under lucerne and 163 kg N/ha under subterranean clover (not significantly different). Rates of net N mineralisation during a subsequent wheat phase (June to December 2001) were 70 kg/ha after lucerne, and 76 kg/ha after the annual legume-based pasture, indicating that lucerne residues, in sandy soils at least, do not mineralise at lower rates compared with subterranean clover residues.

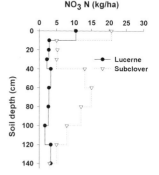

NO$_3$ N (kg/ha)

Figure 3 Amounts of NO$_3$ in soil layers under pastures in May 2000, preceding winter rainfall

Conclusions   The use of lucerne-based pastures did decrease N input through symbiotic N fixation compared with the widely used subterranean clover based pasture. However, the rate of mineralisation of organic N, either during or subsequent to pasture phases, was similar to the traditional annual-based pasture. The perennial growth habit of lucerne ensured quick plant capture of NO_3^{-1} after rain in autumn when land is traditionally bare under annual-based production systems. This capture of NO_3^{-1} by lucerne should reduce the potential for NO_3^{-1} leaching when drainage occurs below the rooting depth attributed to annuals.

References

Bolger T. P., J. F. Angus & M. B. Peoples (2003). Comparison of nitrogen mineralisation patterns from root residues of *Trifolium subterraneum* and *Medicago sativa*. *Biology and Fertility of Soils,* 38, 296-300.

Fillery, I. R. P. (2001). The fate of biologically-fixed nitrogen in dryland farming systems: a review. *Australian Journal Experimental Agriculture,* 41, 361-381.

Characterisation of soil organic matter from Pensacola bahiagrass pastures grazed for four years at different management intensities

J.C.B. Dubeux, Jr.[1], L.E. Sollenberger[2], N.B. Comerford[2], A.C. Ruggieri[3] and K.M. Portier[2]
[1]Av. Dom Manoel de Medeiros, S/N, Dois Irmãos, Brazil, Email: dubeux@ufrpe.br, UFRPE/Dept. Zootecnia, Recife, PE, Brazil CEP 52171-900,[2]P.O. Box 110500, University of Florida, Gainesville, FL, USA 32611-0500, [3]Instituto de Zootecnia, Sertãozinho, SP, Brazil

Keywords: soil organic matter, nutrient cycling, N fertilisation, stocking rate

Introduction Soil fertility and agricultural system sustainability depend upon soil organic matter (SOM), particularly in the tropics, because of highly weathered soils and low fertiliser inputs. Because of the beneficial effects of SOM on chemical, physical, and biological soil properties, Greenland (1994) suggested that SOM is an indicator of agro-ecosystem sustainability. Pasture management may affect SOM by altering the production/decomposition ratio of residues (Johnson, 1995). The objective of this study was to characterise the SOM of Pensacola bahiagrass pastures grazed for four years at a range of management intensities.

Materials and methods Pensacola bahiagrass (*Paspalum notatum*, Flügge) pastures were grazed (2001-2004) at four intensities, defined as the combination of stocking method, N fertiliser, and stocking rate: Continuously Stocked (CS) Low [40 kg N/ha per year and stocking rate of 1.2 animal units/ha (AU = 500 kg live weight)]; CS Moderate (120 kg N and 2.4 AU); CS High (360 kg N and 3.6 AU); and rotationally stocked with 7d grazing period/21d resting period (360 kg N and 3.6 AU). Composite soil samples (0-8 cm) from each pasture were collected in 2004. Particle size distribution was determined on 100 g from each sample by sieving soil into four classes: >250 µm; 150-250 µm; 53-150 µm; <53 µm. SOM density separation was accomplished by decantation and density separation (light and heavy fractions) with water. Physical separation was performed by adapting the methods reported by Meijboom *et al.* (1995), using water instead of Ludox gel. The different class sizes and densities were analysed for C and N by a dry combustion method (Carlo Erba NA-1500 C/N/S analyser).

Results Particle size distribution was not affected by management. Average values were 540, 320, 130, and 10 g/kg for the size classes >250 µm; 150-250 µm; 53-150 µm; < 53µm, respectively. Management intensity altered C and N significantly when particle size was >250 µm. Because this is also the most abundant class size in this Spodisol (540 g/kg), changes have significant impact on SOM. Increased management intensity increased both C and N in both density fractions (Figures 1 and 2). Higher residue deposition at more intensive management, mainly below ground residues (roots + rhizomes) with a lower turnover rate, explained the C and N increases.

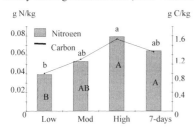
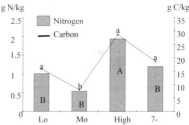

Figure 1 C and N in the light OM fraction > 250 µm[†] **Figure 2** C and N in the heavy OM fraction > 250 µm[†]

([†] For both figures, upper case letters compare values of N and lower case letters compare values of C, within each density class; means with the same letter do not differ (P>0.10) by the LSMEANS test from SAS)

Conclusions Increasing N fertiliser and stocking rate increased C and N in the light and heavy SOM fractions from particles >250 µm. This increase is likely to result in greater C sequestration and soil fertility. Also, the results suggest that changes in SOM concentration can be detected earlier by using the density fractionation technique instead of total SOM determination.

References
Greenland, D. J. (1994). Soil science and sustainable land management, p. 1-15, *In:* J. K. Syers & D. L. Rimmer, (eds). *Soil science and sustainable land management in the Tropics*. CAB International, Wallingford, UK.
Johnson, M. G. (1995). The role of soil management in sequestering soil carbon, p. 351-363, *In:* R. Lal et al., (eds). *Soil Management and Greenhouse Effect*. Lewis Publishers, Boca Raton, FL.
Meijboom, F. W., J. Hassink & M. Van Noordwijk (1995). Density fractionation of soil macroorganic matter using silica suspensions. *Soil Biology and Biochemistry*, 2, 1109-1111.

Organic matter transformation processes of soils in native steppe grass communities

E. Forró

Corvinus University of Budapest, Faculty of Horticultural Sciences, Department of Soil Sciences and Water Management, 1118 Hungary, Budapest, Villányi út 29-43, Email: edit.forro@uni-corvinus.hu

Keywords: soil organic matter, grass plant cover

Introduction It has been well known that higher plant density provides more effective protection for soils. However, the influence of different vegetation on the soil properties has been less well recognised. The high density of grass communities, a high number of plant species and density of roots have advantageous effects on soil properties and conditions, particularly on the organic matter cycling and structure of soils.

Materials and methods Comparative studies are required to show the differences in soils' humus conditions with grass plant cover and the role of the vegetation in the soil protection. In addition to grass vegetation, we investigated soil qualities under *Quercus-, Pinus-, Robinia* forest vegetation, and a cultivated soil with cereals. The habitual physical and chemical properties were investigated. Some properties, which are suitable for the evaluation of humification processes, were also studied. We measured humus quantity (H%), humus quality (Q > 1 – good quality, Q < 1 – weak quality), and calculated the Environmental Protection Capacity (EPC) values of the soil (which was a basic marl) (Hargitai, 1989). EPC is based on the soil's humus status, taking into consideration the humus quantity (H%), the humus quality (Q) and the depth of the humic layer. That is EPC=DxHxQ, where D is the depth of the humic layer (cm) (Forró & Némethy, 1999).

Results Table 1 shows results from a grassland site where many perennial grass species produced a large amount of organic matter and yielded the best quality humus substances (Table 1). An analysis of the processes of the accummulation and decomposition of organic matter showed that the higher was the diversity of plant species, the better was the status or condition of the soils investigated, which was indicated by the advantageous Q value. This allows a higher permissible load of soils, which is also shown by higher EPC values (EPC values of all the examination sites ranged between 35 – 312).

Table 1 Main characteristics of the examination site

Soil depth (cm)	CaCO$_3$ (%)	pH$_{H2O}$	Organic matter (%)	N total (mg/100g)	C/N	**Q**	K	EPC	EPC total
Litter	-	6.43	74	986	44	-	-	-	
Partially decomposed OM	-	7.15	27	941	17	-	-	-	
0-10	2.8	6.89	8	385	12	**1.7**	0.2	65	
10-20	2.3	6.98	5	339	9	**2.9**	0.6	150	**215**
20-30	3.2	7.27	2	277	4	-	-	-	
30-40	3.2	7.51	1	258	2	-	-	-	
40-60	23.5	7.90	1	71	8	-	-	-	

Conclusions We found that with higher species numbers, the complex plant residue spectrum resulted in organic matter accumulation which made the production of stable humic substances possible, thus increasing the potential permissible load of soils.

Acknowledgements This research was supported by the Hungarian Scientific Research Fund (OTKA 34644).

References

Forró, E. & A. Némethy (1999). Distribution of N forms and their changes in soils of natural areas. *10th Nitrogen Workshop Abstracts*, Vol.1, 15.

Hargitai, L. (1989). The role of humus status of soils in binding toxic elements and compounds. *The Science of the Total Environment*, 81/82, 643-651.

Study of soil characteristics to estimate sulphur supply for plant growth

M. Mathot, R. Lambert, B. Toussaint and A. Peeters
Laboratory of Grassland Ecology, Catholic University of Louvain, Place Croix du Sud 5 bte 1, B-1348 Louvain-la-Neuve, Belgium, Email: mathot@ecop.ucl.ac.be

Keywords: sulphur, soil

Introduction During the last decades, sulphur (S) deficiencies have been observed throughout Europe (Zaoh *et al.*, 2002). Accurate estimation of S supply by the soil-atmosphere system is required to give advice for S fertilisers. Soil is an important S source for plants and it is therefore important to evaluate supply by the soil to avoid deficiencies or excessive S fertilisation. The purpose of this preliminary study was to estimate which soil characteristics could be useful for predicting S supply by soil. *Lolium multiflorum* was grown on different soils in a growth chamber and S supply was correlated with soil characteristics.

Materials and methods Twelve soils were sampled at 0-15 cm depth, passed through a 0.5 cm sieve, and air dried at 30° C. Soil DM was determined at 105° C. An equivalent of 1 kg DM was mixed with 250 g of sand, to reduce differences in soil structure; there were 6 replicate pots. Four soils came from permanent grasslands (Pg, soils 1 to 4) and 8 from arable land or temporary grassland (Al, soils 5 to 8). On each pot we sowed 2 g of *Lolium multiflorum* (cv Meroa). Pots were kept at 22°C during the day (16 hours artificial light) and at 16 °C during the night. Each pot received water and all nutrients for optimal growth except S (Lombaert, 1992). Five cuts were harvested and dry matter yields measured. Forages from two replicates were mixed to provide enough sample for analyses. Plant total sulphur content (Leco) and nitrogen content (NIRS) was determined. Soils were analysed for carbon, total S, sulphate content and texture. Plant S yields are the sum of those at each cut.

Results and conclusions Significant differences in S yields were observed between the soils. Soils coming from Pg supplied more S for plants than Al soils (Figure 1) with similar soil characteristics.

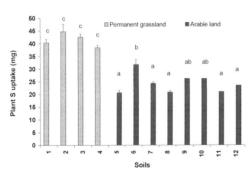

Figure 1 Plant S yields (mean and sd) from soils: different letters indicate a difference at p<0.05 (Systat, 1998)

Table 1 Linear relationship between soil characteristics (x) and S supply estimated by plant S yields (y). (y=a*x+ b: R is the correlation coefficient)

Soil charact.	Pg (n=4)			Al (n=8)		
	a	*b*	*R*	*a*	*b*	*R*
C (g/100g soil)	3.94	32.6	0.893	3.48	17.6	0.781*
S (g/kg soil)	24.3	33.0	0.899	27.0	15.7	0.737*
SO₄ (g S/kg soil)	619.6	36.3	0.833	938.8	18.6	0.426
C/S	-0.12	49.4	0.209	0.24	10.3	0.615
Clay (%)	0.30	36.6	0.755	0.24	20.1	0.453
Silt (%)	0.20	37.4	0.690	0.09	20.8	0.506
Sand (%)	-0.13	49.9	0.751	-0.10	28.8	0.603

* Significant correlation (p<0.05)

Linear correlations were established between soil characteristics and S supply separately for the two groups (Pg *vs* Al, Table 1). Significant (p<0.05) relations were observed for soil C and total S content for Al soils. This relationship was also strong but not significant for Pg, probably because too few soils were studied. For the soil characteristics considered in this study, carbon and total S contents of the soils were the best indicators for estimating S supply by soil.

References

Lombaert, V. (1992). *Micro-cultures méthode Chaminade*. Dossier Agr. D'aspash 5. 1992 p35-51.

Systat (1998). *Systat 8.0 Statistics*. SPSS Science Marketing Department, SPSS Inc., Chicago, USA, 1086 pp.

Zhao, F. J., S. P. McGrath, M. M. Blake-Kalff, A. Link & M. Tucker (2002). Crop responses to sulphur fertilisation in Europe. *Proceedings N° 504, International Fertiliser Society*, York, UK. 28 pp.

Total sulphur content and N:S ratio as indicators for S deficiency in grasses

M. Mathot, R. Lambert, B. Toussaint and A. Peeters
Laboratory of Grassland Ecology, Catholic University of Louvain, Place Croix du Sud 5 bte 1, B-1348 Louvain-la-Neuve, Belgium, Email: mathot@ecop.ucl.ac.be

Keywords: sulphur, plant analysis, N:S ratio, indicator

Introduction Recently, sulphur (S) deficiencies have been observed throughout Europe (Zaoh *et al.*, 2002). Grassland covers about 50 % of the agricultural area in the European Union. In cutting regimes exports are *ca* 30 kg S/ha per year. Atmospheric depositions provide, at the present, *ca*15 kg S/ha per year and S is not considered in fertiliser programmes. Without fertilisation, the S balance at the field scale is negative and therefore some cut swards could be S deficient. Tools for determining whether the grassland is deficient are required. The utilisation of indicators based on S content and N:S ratio was investigated by using S deficient grasses produced in a controlled environment.

Materials and methods Grass (*Lolium multiflorum* cv Meroa) was grown in pots on 12 different soils. The grasses were cultivated in a controlled environment, 16 hours day (artificial light) at 22 °C and 8 hours night at 16 °C. Two treatments, one with S fertiliser, +S, and one without S fertiliser, -S, were followed in 6 replicates. Plants received water and all nutrients for maximum plant growth (Lombaert, 1992). Grasses were considered S deficient when there was a significant difference in DM yield between +S and –S treatments. The grasses were analysed for DM yield, S content (Leco) and N content (NIRS). Total S content and N:S ratio of grasses coming from non deficient (sufficiency) –S treatment were compared with deficient S grasses from the –S treatment.

Results Deficiencies were observed at the second or the third cut depending on the soil. Grasses receiving enough S for maximum plant growth had a S content higher than 2.0 mg S/g DM and a N:S ratio lower than 18.1 (Figure 1). Deficient grasses had a S content lower than 2.3 mg S/g DM and an N:S ratio higher than 17.8.

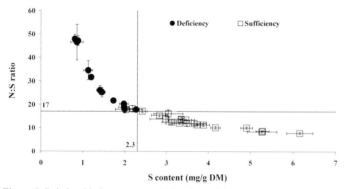

Figure 1 Relationship between grass N:S ratios and S contents (mean and standard deviation) (sufficiency) for S

Conclusions S contents or N:S ratios are useful for determining S deficiency in grasses. Plants with a S content lower than 2.3 mg S/g DM or a N:S ratio greater than 17.0 can be considered as deficient. However, values used as indicators should be adapted to the analytical method used for S determination (Crosland *et al.*, 2001)

References
Crosland, R., F. J. Zhao & S. P. McGrath (2001). Inter-laboratory comparison of sulphur and nitrogen analysis in plant and soils. *Communications in Soil Science and Plant Analysis*, 32, 685-695.
Lombaert, V. (1992). *Micro-cultures méthode Chaminade*. Dossier Agr. D'aspash 5. 1992 p35-51.
Zhao, F. J., S. P. McGrath, M. M. Blake-Kalff, A. Link & M. Tucker (2002). Crop responses to sulphur fertilisation in Europe. *Proceedings N° 504, International Fertiliser Society*, York, UK. 28 pp.

Supplementation of cattle with rock phosphate and urea treated straw to improve manure quality and crop yields in the Sahel zone of Senegal

M. Cissé, M. N'Diaye and C.M. N'Dione
Senegalese Institute of Agricultural Research (ISRA), LNERV, BP 2057, Dakar, Sénégal,
Email: maicisse@refer.sn

Keywords: supplementation, minerals, manure, crop yield

Introduction Mineral deficiencies are a major constraint in improving animal production and crop yield in the Sahel zone (Cissé *et al.*, 1996). Millet (*Pennisetum glaucum*) and groundnut (*Arachis hypogaea*) are two major food and cash crops in this zone. The purpose of this study was to assess effects of supplementing grazing cattle with rock phosphate and nitrogen enriched diets on animal performances, and the effects of the application of their manure on crop yield in a pearl millet-groundnut rotational system located in N Senegal.

Material and methods The study was conducted with 12 farmers. Sixty Gobra cattle, 52 females and 8 males, were equally allotted to a control (group 1) and to 3 other groups which received, during the dry season, concentrates based on phosphorus or/and nitrogen supply after pasture grazing. Cattle received 75 g/animal per day of Thiès rock phosphate in 30 l of water in group 2, 500 g of 4% urea-treated millet stover and 1 kg of peanut cake and 800 g of millet bran/animal per day in group 3, and combining the diet offered in the groups 3 and 2 for group 4. Cattle body condition was scored monthly (Cissé *et al.*, 2003) and manure produced during nights was recorded daily, collected and sun dried. The farm experiment was a millet (*var.* souna 3) groundnut (*var.* Fleur 11) rotational cropping system with 5 treatments: control (no manure), manure from unsupplemented animals (group 1), and manure from cattle group 2, 3, and 4, respectively. During the rainy season, manure was applied at 4 t/ha to millet. Groundnut was planted in the following year without renewing manure application. Plant growth and yield were measured at 24, 52 days and at harvest.

Results and discussion There were important changes in body condition score (BCS) according to the supplement given to cattle. Controls lost (p<0.05) 0.9 points in BCS (3.6 *vs* 2.7) while cattle supplemented with rock phosphate mixed in water maintained their BCS at 3.5 points. Animals from groups 3 and 4 gained (p<0.01), respectively, 0.7 (2.8 *vs* 3.5) and 0.9 points (3.1 *vs* 4) of BCS; this being in part due to the high energy content of their diet. After 28 days' growth and at harvest, millet and groundnut plant populations were not significantly influenced by manure application. At 52 days, manured plants were slightly taller than the controls. Enriching manure resulted in positive response in groundnut plant leaf number and height (Table 1). Millet grain yield increased by 24 to 68% depending on the diet offered to animals. The control without manure provided the lowest yield and the highest production was obtained with additional supply of P and N by manure. However, compared with the production from plots manured by control animals, the gain in millet grain yield due to manure enriched in P and N (i.e., 264 kg/ha) was higher than the sum of the gains due to supplementation either in P (73 kg/ha) or in N only (92 kg/ha). The residual effect of manure on groundnut yield resulted in an increase of 11 to 25% over the yield from the unmanured plots.

Table 1 Effect of manure on number of leaves/plant 52 days after planting, on plant height and on grain yield

Treatment	Direct effect on millet		Residual effect on groundnut			
	Grain yield, kg/ha	% increase	Number of leaves	Plant height, cm	Grain yield, kg / ha	% increase
Control	599c	-	51.93a	19.09a	683b	-
Manure from group 1	744b	24	51.80a	20.79c	742ab	9
Manure from group 2	817b	36	52.56a	20.16b	756ab	11
Manure from group 3	836b	39	57.55b	20.27b	842ab	23
Manure from group 4	1008a	68	58.75b	21.24d	857a	25

Means followed by different letters in the same column are different at p< 0.05

Conclusions This trial demonstrated advantages of supplementation. However, better response in crop yields could be gained if animals were stabled in fields, because of increases in nutrient cycling from faeces and urine.

References
Cissé M., H. Guérin & E. Prince (1996). Les carences minérales existent au Sénégal. Comment corriger ce déficit nutritionnel en élevage? Etudes et documents, *ISRA (ed.), vol.7, no. 1, 33 p.*
Cissé M., A. Korréa, I. Ly & D. Richard (2003). Change in body condition of zebu cattle under different level of feeding. Relationship with body lipids and energy. *Journal of Animal Feed Science, 12, 485-495.*

Nitrogen response of spring and winter wheat to biosolids compared to chemical fertiliser

W. Kato[1,2], O.T. Carton[1], D. McGrath[1], H. Tunney[1], W.E. Murphy[1] and P. O'Toole[2]
[1]Teagasc, Johnstown Castle Research Centre, Wexford, Ireland, Email: kato@vmas.kitasato-u.ac.jp, [2]University College Dublin, Dublin 4, Ireland,

Keywords: biosolids, nitrogen response, spring wheat, winter wheat

Introduction Irish sewage sludge production was over 30,000 t/year in the 1990s (EPA, Ireland, 2003). Application to agricultural land is a management option for this organic material as it results in the recycling of the nutrients they contain for crop production. The EU Directive (91/271/EEC) encourages the recycling of sewage sludge as biosolids to agriculture. However, up to 1999, only about 5 % of biosolids produced was applied to agricultural land. In this study, several biosolids and a chemical fertiliser were used to assess N availability for spring and winter wheat (*Triticum aestivum*) production in a pot experiment.

Materials and methods The experiments were carried out in a solarium from May to July 2001 for spring wheat, and from December 2001 to June 2002 for winter wheat. Three types of biosolids [anaerobic biosolid (AB), dried biosolid (DB) and lime biosolid (LB)], cattle slurry (CS) and chemical fertiliser (CF) were used as N sources. The materials were applied at rates of 90 and 180 mg N/pot for spring wheat and 180 and 360 mg N/pot for winter wheat. There was also a control treatment with zero-N in each case. P and K were applied at sowing time to meet crop requirements. In each pot (area 227 cm^2) 2 kg of loam shale soil (87.5 % dry matter, pH 5.9) was placed over 1 kg of sand. The wheat was sown at a rate of 1.0 g/pot and harvested at the vegetative stage. The dry weight and N content were measured.

Results and discussion The N uptake (mg/pot) for all the treatments and the equations for the linear responses to CF in spring and winter wheat are shown in Figure 1. The N uptakes from AB, LB and CS (90 and 180 mg N/pot) in spring wheat were not much different from each other, however, CF was the highest and DB was significantly the lowest. For example, in the 180 mg N/pot treatment were CF 198.3 (a), AB 115.4 (b), LB 102.0 (b,c), CS 88.1 (c) and DB 31.6 (d) mg/pot (means with a letter in common are not significantly different). In contrast, there were some different trends between the treatments in N uptake for winter wheat (180 and

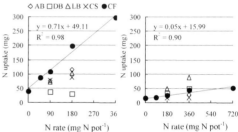

Figure 1 N uptake (mg per pot) for spring (left) and winter wheat (right)

360 mg N/pot); the CF treatment gave a lower slope than in spring wheat, in addition LB and DB gave a higher yield than CF. For example, in the 180 mg N/pot treatment N uptakes were LB 49.3 (a), DB 33.2 (b), CF 24.8 (c), AB 22.2 (c) and CS 17.2 (c) mg/pot. From the above, the efficiency (E) of N uptake in the biosolids relative to CF (Table 1) was calculated by the following formula (Pommel, 1995): E (%) = (A$_1$/A$_2$)×100, where A$_1$ = slope for biosolids and A$_2$ = slope for CF.

The relative efficiency of each biosolid in spring wheat was 48, 39 and 30 % for AB, LB and CS, respectively, while DB was negative. In contrast, LB and DB showed higher responses (over 100 %) in winter wheat. This may indicate that calcium or organic bound N in LB and DB was converted to more available forms before or during winter wheat growth.

Table 1 Relative efficiency (%) of each biosolid as an N source compared to chemical fertiliser for N uptake

	AB	DB	LB	CS	CF
Spring wheat	48	<0	39	30	100
Winter wheat	45	116	249	15	100

Conclusions Biosolids can be used to replace part of the N requirements of wheat. The relative efficiency of LB and DB for winter wheat is higher than CF, however, biosolid N for spring wheat is generally lower than in CF. Biosolids cannot be relied upon to supply sufficient N to produce full crops of spring wheat unless applied the previous year. Therefore, supplementary fertiliser N should be used with this crop.

References

EPA, Ireland (2003) Urban Waste Water Discharges in Ireland, A Report for the Years 2000/2001. Environmental Protection Agency, Ireland. 99pp.

Pommel, B. (1995) Value of a heat-treated sludge in the phosphorus fertilization. *European Journal of Agronomy* 4, 395-400.

Improving nutrient supply of grassland soil

G. Füleky[1] and M. Orbán[1,2]

[1]*Szent István University, Department of Soil Science and Agricultural Chemistry, Gödöllő, Hungary, H-2103 Páter Károly u. 1, Email: Fuleky.Gyorgy@mkk.szie.hu.*, [2]*LAM Alapítvány, Illyefalva, 4016 Romania*

Keywords: fertilisers, grassland, nutrient uptake

Introduction In the Southern corner of the Eastern Carpathian Mountains is located the hilly region of Barcaság. The total agricultural area is 180 thousand ha of which 90 thousand ha is grassland. The maximum yield of natural grassland does not exceed 1.2 t ha[-1] hay. The aim of this work is to find appropriate fertiliser and liming rates to increase the hay yield and improve the soil nutrient supply.

Methods A fertiliser and lime field trial was established with three replicates in a typical valley area of this region on Fluvisol in 2001 (Orbán *et al*., 2002). The average precipitation is 753 mm, the average temperature 7.3 °C. pH_{KCl} of the soil is 5.67, humus content 2.24 %, the texture clay-loam, and the soil has medium available P and very low K contents. The lime and fertiliser rates can be seen in Table 1. Ca and N, P, K were supplied as sugar beet factory lime, NH_4NO_3, superphosphate and KCl, respectively. A mixture of grasses and legumes was sown. The floristic composition of swards were determined twice. The dry matter yield and the chemical composition of the hay was also determined.

Results The floristic composition of plants was different in the two years. In the first year the ratio of grasses was about 25 % on the plots not fertilised with nitrogen (N). However, on the plots fertilised with N this ratio was about 50 %. The ratio of legumes was highest on the plots not fertilised with N. With increasing nutrient supply the ratio of legumes decreased to 30 %. In the second year, the ratio of grasses was *ca* 75-80 %, and that of legumes 10 %. The negative relationship between the ratio of legumes and the nutrient supply of soil was significant in the first year. Table 1 shows the cumulative yields and N, P and K uptake. As the results demonstrate, liming alone was not satisfactory to improve soil fertility and combining it with N, P, and K fertilisers significantly increased the hay yield. As the balance calculations show, the applied N_{50} and N_{100} rates were not enough to balance the N uptake of plants. By using 60 kg P for 3 years the P supply of soil did not decrease. It is surprising to see that the K rate of 240 kg for 3 years was not sufficient to maintain the K supply of soil.

Table 1 Dry matter yield and macro element uptake of grassland plants (sum of results from 2002 and 2003)

Treatment	Yield	N kg/ha		P kg/ha		K kg/ha	
N kg/ha per yr P, K kg/ha per 3 yr Ca 2t/ha per 3 yr	t/ha	Uptake	Balance	Uptake	Balance	Uptake	Balance
Control	9.0	197	-197	31	-31	248	-248
Ca	9.2	173	-173	29	-29	200	-200
CaN_{50}	11.0	213	-113	35	-35	277	-277
$CaN_{50}P_{60}$	11.9	256	-156	38	+22	306	-306
$CaN_{50}P_{60}K_{240}$	12.0	228	-118	34	+26	335	-195
CaN_{100}	12.9	266	-66	40	-40	339	-339
$CaN_{100}P_{60}$	12.6	260	-60	42	+18	351	-351
$CaN_{100}P_{60}N_{240}$	13.9	291	-91	43	+17	424	-184
LSD 5%	0.5	23		4		32	

Conclusions The limited hay production of poor soils could be increased with liming and fertilisers. The improved nutrient supply of soils was demonstrated with increased macro element uptake.

References

Orbán M., G. Füleky & I. Razec (2002). Correlation between soil and hay quality of natural grassland in Barcaság hilly region. *Innovation, science and practice of agriculture at the millennium* (In Hungarian), Debrecen 11-12 April 2002. p. 172-177.

Section 3

Physical constraints to soil formation

Assessment of nitrogen nutrition status of grasses under water deficit and recovery

V.G. Dugo, J-L. Durand and F. Gastal
Ecophysiologie des Plantes Fourragères, INRA, 86 600 Lusignan, France, Email: jldurand@lusignan.inra.fr

Keywords: tall fescue, Italian ryegrass, nitrogen, water potential

Introduction Grasslands are rarely irrigated. They are therefore systematically submitted to more or less severe water deficits: as well as mineral deficiencies, water scarcity often also results in a reduction of nitrogen (N) status. Although identified some time ago, qualitatively, the interaction with N still remains difficult to take into account in quantitative analyses of crop physiology under water deficits. This paper illustrates how the nitrogen (N) status of the crop changes under water deficits. A N nutrition index (INN) was defined as the ratio of the actual N concentration of forage with the theoretical N concentration under optimal conditions, the latter only depending on the above ground biomass. The objective of the paper is to describe the effect of water deficits on INN, using a new assay recently proposed by Faruggia *et al.* (2004).

Materials and methods Two grass species (*Lolium multiflorum* (*Lm*) and *Festuca arundinacea* (*Fa*)) were grown in swards and either sheltered from rainfall or fully irrigated to match potential evapotranspiration, and at two levels of N fertiliser. Soil humidity (neutron probe), soil N (organic and mineral) content, pre-dawn leaf water potential (Ψ_D) were measured during one summer re-growth in 2003 in Lusignan, France, 46.43 ° N, 0.12° W. Nitrogen concentration of upper leaves (N_{sup}) was measured as an assessment of INN according to the procedure described in Farrugia *et al.* (2004). Dry herbage and root mass extracted from 25 cm long x 8 cm diameter soil cores sampled to 4 cm depth intervals were measured at beginning and end of the study period.

Results Following a dry spring and in spite of an initial irrigation on 18 June, both species exhibited a rapid decline in Ψ_D from -0.2 MPa down to -1.4 MPa on 28 June. A full recovery in water status was observed in both species at both N fertiliser levels after water had been given on all plots on 30 June. On all occasions, *Fa* appeared to capture mineral N much more slowly than *Lm* both under dry conditions after cutting and during recovery (Figure 1). The growth rate of *Lm* was much lower than that of *Fa*, partly explaining the differences in the ability to maintain a high INN at low Ψ_D. The root mass distribution was shallower in *Lm* than in *Fa* and this is also likely contributing to the easier recovery of N from soil in the former species. The INN results were consistent with mineral N and water distribution in soil.

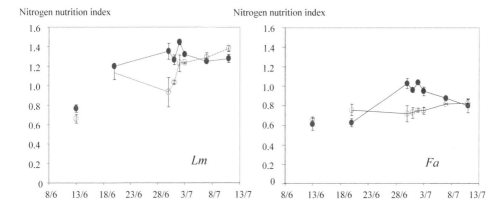

Figure 1 Time course of nitrogen nutrition index during drought and recovery in high N treatments; closed circles = irrigated and open circles = dry treatments

Conclusions The experiment showed that the new assessment tecnique of INN can be used to follow the short term impact of the sward's water status on plant N status. However, root characteristics for water and N uptake might be different and further studies are required to fully ascribe the different capacities for maintaining high INN at low water status for each species.

References

Farruggia A, F. Gastal & D. Scholefield (2004). Assessment of the nitrogen status of grassland. *Grass and Forage Science*, 59, 113-120.

Denitrification under pastures on permeable soils helps protect ground water quality

M.P. Russelle[1], B.A. Browne[2], N.B. Turyk[2] and B. Pearson[2]
[1]USDA-Agricultural Research Service-Plant Science Research Unit, 1991 Upper Buford Circle, Room 439, University of Minnesota, St. Paul, Minnesota 55108-6028, USA, Email: russelle@umn.edu, 2College of Natural Resources, University of Wisconsin, 800 Reserve Street, Stevens Point, Wisconsin 54481, USA

Keywords: soil nitrate, ground water, nitrate leaching, dissolved organic carbon, nitrous oxide

Introduction Pastures have been implicated in ground water contamination by nitrate, especially in humid regions with thin or sandy soils (Stout et al., 2000). Significant losses can occur even under low N input, because available N from excreta patches often exceeds plant uptake capacity. Lack of evidence that appreciable nitrate leaching was occurring in established Midwestern USA pastures led us to test the hypothesis that denitrification was preventing or remediating nitrate loading. Higher denitrification rates have been found in the relatively limited number of trials since Ball & Ryden (1984) first reported the significance of this process in pastures.

Materials and methods At three grazing dairy farms located on soils with high hydraulic conductivity in central Wisconsin, USA, multiport ground water wells were established on the up-gradient and down-gradient edges of at least one paddock and in a field under corn-soybean management on a nearby confinement dairy farm. Inorganic N was determined in the upper 1.2m of soil. In one paddock and in the corn-soybean field, an intensive grid of mini-piezometers was established to determine the range of variation in *in situ* dissolved solids and gases in ground water, sampled by pumping-induced ebullition (Browne, 2004). Two independent experiments were conducted in a growth chamber on intact soil cores (6-cm diam. by ~60-cm long) from one paddock, with or without fresh dairy cow excreta applied at the start of each 28- to 31-d incubation period.

Results Ground water samples from the multiport wells indicated that nitrate was leaching at substantially smaller rates than under other agricultural practices in the area. Although differences in soil nitrate concentration were evident between excreta spots and background areas on several sampling dates, no differences in dissolved organic carbon were detected. The intact soil core experiments provided convincing evidence that urine increased denitrification. Soil pH, ammonium, and nitrate concentrations followed patterns reported by others, the first two increasing rapidly after urine application, and nitrate increasing after about 7 days. Smaller changes occurred under fresh dung than fresh urine. Nitrous oxide emission over 4 weeks was 3-fold higher with dung and 9-fold higher with urine than the control soil. Methane emission was 20-fold higher with dung than either urine or no treatment, whereas CO_2 emission quadrupled with either excreta. In the field, there was tremendous spatial variability in ground water chemistry (dissolved nutrients and gases). Figure 1 shows that the dissolved denitrified N (measured as dissolved N_2 gas in excess of atmospheric N_2) was higher as a percentage of total nitrate (nitrate + denitrified N) in groundwater beneath the pasture (n >60 sites per sampling time) than beneath the arable field (n >20 per sampling time), and ancillary measurements (e.g., dissolved organic C and dissolved O_2) supported this result. In contrast, dissolved N_2O was lower under the pasture than corn. Further research is underway to determine the variation in dissolved gas concentrations in ground water under other grazed paddocks and arable fields, the sources of the dissolved organic carbon, the proportion of N lost as N_2 in these systems, and the potential of denitrification to reduce nitrate loading of ground water in the region.

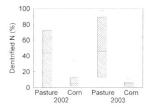

Figure 1 Denitrified N (excess N_2 as a percent of nitrate+denitrified N) in ground water [percentiles of data are shown by whiskers (10th and 90th), boxes (25th and 75th), and horizontal line (50th)]

Conclusions This is the first report of the wide variation in dissolved gas composition in ground water under pastures. The field evidence and results from intact soil cores lend support to the hypothesis that denitrification may remove substantial amounts of N from pastures. Enhanced denitrification may benefit water quality more generally as ground water moves from or toward adjacent arable cropland.

References
Ball, P. R. & J. C. Ryden (1984). Nitrogen relationships in intensively managed temperate grasslands. *Plant and Soil*, 76, 23-33.
Browne, B. A. (2004). Pumping-induced ebullition: A unified and simplified method for measuring multiple dissolved gases. *Environmental Science and Technology*, 38, 5729-5736.
Stout, W. L., S. L. Fales, L. D. Muller, R. R. Schnabel, & S. R. Weaver (2000). Water quality implications of nitrate leaching from intensively grazed pasture swards in the northeast USA. *Agriculture, Ecosystems, and Environment*, 77, 203-210.

Phosphorus transfer to river water from grassland catchments in Ireland

H. Tunney[1], P. Jordan[2,] G. Kiely[3], R. Moles[4], G. Morgan[3], P. Byrne[4], W. Menary[2] and K. Daly[1]
[1]Teagasc, Johnstown Castle, Wexford, Ireland, Email: htunney@johnstown.teagasc.ie, [2]University of Ulster, Coleraine, BT52 ISA, UK, [3]University College, Cork, Ireland, [4]University of Limerick, Ireland

Keywords: phosphorus, transfer, water, grassland

Introduction In Ireland it is estimated that at least half of phosphorus (P) loss to water is from agricultural sources and National and European Union policy and legislation aim at reducing phosphorus (P) loss to water in order to reduce eutrophication. In Ireland, the average soil test P (STP) levels increased ten-fold, from less than 1 to over 8 mg Morgan P per 1 soil over the past 50 years, reflecting increased P inputs in fertiliser and animal feed. One of the main objectives of this three-year research programme, started in 2001, was to investigate P loss to water in grassland catchments.

Materials and methods Phosphorus loss to water was studied in three catchments (with nested sub-catchments), one in the north (Oona, Co. Tyrone; shale soil), centre (Clarianna, Co. Tipperary; limestone soil) and south (Dripsey, Co Cork; old red sandstone soil) of the island. This involved setting up field stations for the collection of hydrological and water chemistry data in the nested catchments at different scales (Table 1), investigating the loss of different P fractions and suspended solids in the river water (runoff) under various seasonal, meteorological, hydrological and soil conditions. The mean intensity of grassland farming and STP were broadly similar in the smaller subcatchments of the three catchments.

Results The differences in hydrology, rainfall and soil types between the three catchments were reflected in runoff and P export to water.The Dripsey and Oona catchments had broadly similar total P (TP) exports, of the order of 2 kg P/ha per year (Table 1). This level of loss is higher than the level of about 0.5 kg TP/ha that is considered compatible with good water quality. In general Oona had higher SS than the other two catchments, probably reflecting soil type and more intensive runoff. In contrast, Clarianna had a several fold lower P export per unit land area despite having broadly similar STP levels and agricultural intensity as Dripsey and Oona. The Clarianna had less runoff and has mainly thick calcareous Quaternary deposits which retain P more effectively than the other two catchments. Losses of P per unit area were influenced by catchment size and STP.

Table 1 Rainfall, evapotranspiration (ET), catchment area, runoff, and mean values in the river water for P fractions and corresponding loss per ha (load) for total P (TP), particulate P (PP) and dissolved reactive P (DRP) for the three catchments from 1 Jan. to 31 Dec. 2002. The error statistics (%E) were calculated for loads from the 95% confidence limit least squares regression equations used to gap fill time series water chemistry data

Basin	Rain	ET	Area	Runoff	TP			PP			DRP		
	mm	mm	km²	mm	mg/l	kg/ha	%E	mg/l	kg/ha	%E	mg/l	kg/ha	%E
Dripsey	1833	362	0.17	1206	0.22	2.66		0.049	0.60		0.15	1.85	
			2.11	1080	0.23	2.48		0.099	1.07		0.11	1.14	
			14	1037	0.15	1.60		0.057	0.59		0.08	0.81	
Oona	1366	352	0.15	611	0.39	2.40	2.89	0.239	1.46	3.57	0.08	0.51	3.45
			0.62	894	0.27	2.41	2.07	0.157	1.40	3.57	0.06	0.52	3.57
			88.5	817	0.38	3.13	6.40	0.203	1.66	5.77	0.11	0.90	4.62
Clari-anna	1091	493	0.8	603	0.11	0.69	3.77	0.078	0.47	3.62	0.02	0.15	4.00
			7.3	435	0.07	0.30	1.67	0.049	0.21	1.43	0.01	0.06	5.00
			13.6	416	0.04	0.17	5.88	0.021	0.09	6.67	0.01	0.06	8.33
			29.8	434	0.05	0.23	8.26	0.021	0.09	10.0	0.03	0.11	6.36

Conclusions Hydrology and soils in some catchments (e.g. Clarianna) can minimise the loss of P compared with other catchments (Oona and Dripsey) with broadly similar mean STP and grassland farming practices and these factors are important determinants in P transfer from catchments. The relative importance of factors influencing P transfer from grassland to water will help in agreeing the most appropriate management practices to help reduce loss to water.

Maximising slurry crop available nitrogen utilisation in grassland systems

J.R. Williams[1], E. Sagoo[1], B.J. Chambers[2], J. Laws[3] and D.R. Chadwick[3]

[1]*ADAS Boxworth, Battlegate Road, Boxworth, Cambridge, CB3 8NN, UK, Email: john.williams@adas.co.uk,* [2]*ADAS Gleadthorpe, Meden Vale, Mansfield, Notts. NG20 9PF UK,* [3]*IGER North Wyke, Okehampton, Devon, EX20 2SB UK*

Keywords: ammonia volatilisation, nitrate leaching, slurry application timing

Introduction In the UK, approximately 90 million tonnes of animal manure containing *ca* 450,000 tonnes of nitrogen (N) are recycled to agricultural land each year. The efficient utilisation of manure N can save farmers money and reduce diffuse air (ammonia) and water (nitrate) pollution. For slurries, bandspreading techniques (e.g. trailing shoe and trailing hose) can improve N utilisation by reducing ammonia volatilisation losses compared with conventional broadcast applications. They also provide increased spreading opportunities in spring/summer as slurry is placed in a band on the soil surface limiting herbage contamination, which can reduce the need to apply slurry in the autumn/early winter period that can exacerbate nitrate leaching losses. However, spring/summer application timings (when temperatures are higher and soils are drier) may lead to increased ammonia emissions compared with autumn/winter applications under cooler and moister soil conditions. This paper reports results from a project to investigate the effects of contrasting slurry application timings on ammonia volatilisation and nitrate leaching losses and grass N utilisation.

Methodology Experiments were set up on two commercial dairy farms; Betley (Cheshire) and Inkberrow (Worcestershire) in autumn 2002. Ammonia emissions and nitrate leaching losses (autumn timing at Betley) and crop dry matter yields and N uptake were measured following different slurry application timings (Table 1). There were 3 replicates of each slurry application timing along with accompanying fertiliser N response treatments (0-250 kg/ha N) to quantify the fertiliser N replacement value of the slurry dressings.

Table 1 Slurry application timings

Site	Application method	Application timing
Betley (Cheshire)	Tanker with 12m trailing hose boom	October 2002, February, March and April 2003 (first cut)
		Early June and late June 2003 (second cut)
Inkberrow (Worcestershire)	Umbilical system with 6m trailing shoe boom	February, March and April 2003 (first cut)

Results and discussion At Betley, ammonia emissions were highest ($P<0.05$) following the slurry application in early June (before second cut) at 17% of the total N applied and lowest following the October timing at 4% of the total N applied. The higher ammonia losses following the early June timing were most probably due to a combination of higher soil temperatures (15°C), 'dry' soil conditions and lack of grass cover (<5cm height) compared with the other application timings (soil temperature range 1-10°C and grass heights > 7.5 cm). Nitrate leaching losses following the October application were not different from the untreated control ($P>0.05$), indicating that slurry N had been taken up over winter by the grass sward. At both first and second cut, there was, surprisingly, no response ($P>0.05$) to either fertiliser or slurry N additions. At Inkberrow, ammonia emissions were similar ($P>0.05$) following each of the three spring application timings at between 6 and 8% of the total N applied. The fertiliser N replacement value of the slurry applications (based on crop N offtake) was 70 kg/ha, 63 kg/ha and 54 kg/ha for the February (91 days before cutting), March (63 days before cutting) and April applications (43 days before cutting), respectively. The number of days slurry could be spread was estimated to increase from 55 days/year for surface broadcasting to 145 days/year with the trailing shoe (bandspreading) machine. This was mainly because reduced herbage contamination allowed grazing to occur within a few days of application (rather than waiting 3-4 weeks following surface broadcasting) and a greater spreading window before silage cutting (rather than finishing applications 6-8 weeks before cutting with surface broadcasting).

Conclusions There is a need to ensure that slurry management policies that aim to reduce nitrate leaching (i.e. moving from autumn application timings to late spring/early summer) do not exacerbate ammonia losses through increased emissions under warmer and drier soil conditions. An integrated approach to slurry N management is needed that considers all N loss pathways and aims to maximise crop N utilisation.

Fire and nutrient cycling in shortgrass steppe of the southern Great Plains, USA

P.L. Ford[1] and C.S. White[2]
[1]USDA Forest Service, Rocky Mountain Research Station, Albuquerque, New Mexico, Email: plford@fs.fed.us,
[2]University of New Mexico, Albuquerque, New Mexico, USA

Keywords: nutrient cycling, fire, shortgrass steppe, potentially mineralisable nitrogen, plant cover

Introduction Fire in semi-arid grasslands releases nutrients bound up in organic matter and accelerates the rate of decomposition in the soil. This research experimentally tested effects of season and frequency of fire on nutrient cycling dynamics in shortgrass steppe. The objective was to identify if fire treatments have the ability to increase potential grassland productivity relative to untreated 'reference condition' grassland. Many such studies focus on short-term, direct effects of fire. However, this study is part of a long-term, 18-year study examining both direct, and indirect effects of fire in the growing *vs.* dormant season at return intervals of 3, 6 and 9 years.

Materials and methods The study is located in semi-arid shortgrass steppe in the southern Great Plains of northeastern New Mexico, USA (36° 31' 20" N, 103° 3' 30" W). The never-ploughed, ungrazed, 160-ha site has mostly native vegetation with the sod-forming *Buchloë dactyloides* and the bunchgrass *Bouteloua gracilis* being the dominant plant cover. The experimental design was completely randomised with 5 treatments and 5 replicate 2-ha plots per treatment. Treatments were 3–year dormant- (3D) and growing-season (3G) burn cycles (twice burned), 6-year dormant- (6D) and growing-season (6G) burn cycles (burned once), and unburned, reference condition (RC) plots. The first two rounds of treatments were applied in 1997 and 2000 with measurements taken during the drought year of 2003. Response variables included % bare ground, litter and live perennial grass cover, soil organic matter content, pH, sodium adsorption ration, field available nitrogen (N) as nitrate and ammonium, potentially mineralisable N (PMN), and other soil and plant nutrients.

Results The only suggested difference in cover variables among treatments (p = 0.07) was between the unburned (RC) plots with a mean of 60% litter cover and the 3-year dormant-season (3D) fire treatment with a mean of 46% litter cover (Figure 1, Table 1). There were significant differences in the levels of boron, calcium, and sodium in vegetation among treatments, but no generalities could be made regarding fire effects. Depending on the frequency, fire can either increase or deplete soil nutrients. The main differences among soil variables occurred between the 6D and 3D fire treatments (Table 1). When differences were suggested, the 6D treatment always had a significantly higher mean than 3D (p = 0.06), but did not significantly differ from the other treatments (Table 1 and Figure 1). In addition, ammonium was significantly higher in 6D than in RC.

Table 1 Tukey's Studentised Range Test means (values the same letter are not significantly different (alpha = 0.10))

Grouping		Mean (SD)	n	Treatment
Litter %	A	60 (1)	5	RC
	AB	54 (5)	5	6G
	AB	54 (8)	5	6D
	AB	49 (9)	5	3G
	B	46 (10)	5	3D
PMN:	A	41 (8)	5	6D
Nitrate	AB	36 (6)	5	RC
µg/kg	AB	35 (6)	5	6G
	AB	32 (4)	5	3G
	B	30 (2)	5	3D
PMN:	A	0.70 (.09)	5	6D
Ammonium	AB	0.52 (.13)	5	3G
µg/kg	AB	0.51 (.07)	5	6G
	B	0.49 (.19)	5	RC
	B	0.47 (.10)	5	3D

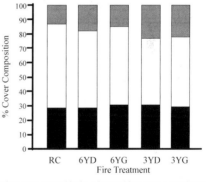

Live grass cover (black), Litter (white), Bare ground (grey)

Figure 1 Ground cover by treatment

Conclusions The current results of this long-term study suggest that in semi-arid grasslands a 3-year fire frequency (burned twice in 6 years), regardless of season, may be too short, and may cause a greater loss of litter and limiting N resources, than other frequencies. The 6-year dormant-season fire (i.e. burned once in 6 years), is the only fire treatment that shows the potential for increased site production relative to 'reference condition' unburned grassland.

Soil aggregate dynamics, particulate organic matter and phosphate under dryland and irrigated pasture

J.T. Scott[1], L.M. Condron[1] and R.W. McDowell[2]

[1]AgResearch, c/-Agriculture and Life Sciences, PO Box 84, Lincoln University, New Zealand, Email: john.scott@agresearch.co.nz, [2]AgResearch, Invermay Agricultural Centre, Private Bag 50034, Mosgiel, New Zealand

Keywords: soil aggregates, phosphate, particulate organic matter, irrigation

Introduction Soil aggregate formation and turnover affects the rate of occlusion or release of soil organic matter and therefore the availability for mineralisation or stabilisation of soil carbon (C) and phosphorus (P). Furthermore, differences in soil type, management and the quantity and quality of organic inputs can affect aggregate turnover rates (Six et al., 2000). Under pastoral farming the ratio of coarse particulate organic matter (inter-POM) inside macroaggregates but outside microaggregates to fine POM (intra-POM) within microaggregates may provide an indication of physical processes influencing mineralisation and stabilisation of soil C and organic P (Po). Our aim was to determine the coarse and fine POM and associated C and P contents in water stable macro and microaggregates under long term irrigated and dryland pasture grazed by sheep.

Materials and methods Soil to 75 mm depth was collected from irrigated and dryland pasture under sheep grazing at a long term irrigation trial site at Winchmore in New Zealand. Field moist soil was gently sieved to <2 mm, air dried then wet sieved by hand to obtain water stable 2000 - 2500 μm macroaggregates. Macroaggregates were broken up by using the method of Six et al. (2000) to obtain inter-POM and intra-POM and to determine the sand free content of both macro and microaggregates. Respective soil aggregate and POM fraction weights, total P (TP), inorganic P (Pi), Po and C content were determined.

Results Greater sand free microaggregate weight was obtained in irrigated than unirrigated soils. Olsen P levels were not significantly different at 31 μg/ml in complete soils under the two treatments. The similar proportion of microaggregates in macroaggregates under irrigated and dryland pastures suggests a similar aggregate turnover rate and therefore exposure of POM to mineralisation. However, the ratio of intra- to inter-POM (Table 1) was almost three times greater in irrigated than dryland which suggests slower macroaggregate turnover under irrigation. Irrigated pasture supports a greater worm biomass than dryland and earthworms have been shown to provide protection of soil C in microaggregates explaining the greater proportion of protected POM under irrigation. The greater inter-POM under dryland with a concomitant greater quantity of TP, Pi and Po than under irrigation (Table 1) indicates more POM and P potentially available for plant uptake.

Table 1 Particulate organic matter (POM) and aggregate relationships under dryland and irrigation conditions

Treatment	Intra-/inter-POM	Microaggregates in macroaggregates (%)	Inter-microaggregate POM (g C/ kg macroaggregates)	Intra-microaggregate POM (g C/kg macroaggregates)
dryland	4.81	64.5	1.45	5.47
irrigated	13.63	69.7	0.79	4.37
LSD$_{0.05}$	3.65	6.0 ns	0.66	1.40 ns

Table 2 Phosphorus content of inter-microaggregate POM under dryland and irrigation

Treatment	Total P within inter-POM (μg P/g inter-POM)	Pi within inter-POM (μg P/g inter-POM)	Po within inter-POM (μg P/g inter-POM)
dryland	1465	659	807
irrigated	1239	439	799
LSD$_{0.05}$	173	70	175 ns

Conclusions Although an apparent similar aggregate turnover rate with and without irrigation, the greater intra- to inter-POM ratio and TP, Pi and Po without irrigation indicates more POM and P not being utilised or mineralised because of a lack of biological activity through insufficient moisture.

References

Six, J., E. T. Elliot & K. Paustian (2000). Soil macroaggregate turnover and microaggregate formation: a mechanism for C sequestration under no-tillage agriculture. Soil Biology and Biochemistry, 32, 2099-2103.

Fine colloids 'carry' diffuse water contaminants from grasslands

P.M. Haygarth[1] and A.L. Heathwaite[2]

[1]Institute of Grassland and Environmental Research Station (IGER), North Wyke Research Station, Okehampton, Devon, EX20 2SB, UK, Email: phil.haygarth@bbsrc.ac.uk, [2]Centre for Sustainable Water Management, The Lancaster Environment Centre, Lancaster University, Lancaster, LA1 4Y, UK

Keyword: phosphorus, colloids, hydrology, pathways

Introduction The transport of diffuse pollutants from grassland has traditionally been described by the operationally defined threshold of greater, or smaller than a nominated membrane filter size. Most commonly this has been a 0.45 μm threshold to define 'solute' and 'particulate' transport. In this paper we shall use phosphorus (P) to help provide an example of the importance of colloid-facilitated transport.

Materials and methods Phosphorus transport from grassland soils in different hydrological pathways was investigated using a series of laboratory and field experiments (for fuller details see Heathwaite *et al.*, 2005). In the first instance, a simple laboratory shaking 'batch test' was developed, to provide preliminary information on the propensity of different soils to release P attached to soil colloids. In the second part of the work, the relative contribution of different particle size fractions in transporting different P forms in agricultural runoff from grassland soils was evaluated using a randomised plot experiment, involving various types of P amendment and hydrological pathways.

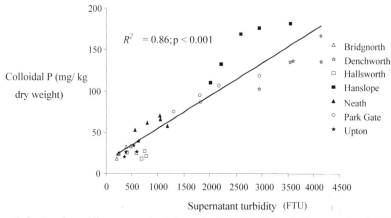

Figure 1 Scatter plot and linear regression between 'colloidal P' (H_2O - $CaCl_2$ extracts) and turbidity for a range of soils subjected to a 'batch test' (the legends refer to different soil series under the England and Wales system)

Results and discussion In the colloidal P 'batch test', the relationship between turbidity of soil extracts and total P (TP) was significant ($r^2=0.996$, $p<0.001$) across a range of grassland soils, and a strong positive relationship ($r^2= 0.86$, $p<0.001$) was found between 'colloidal P' (H_2O - $CaCl_2$ extracts) and turbidity (Figure 1). Linear regression of the proportion of fine clay (<2 μm) for each soil type evaluated against the (H_2O-$CaCl_2$) colloidal P fraction gave a weak but positive relationship ($r^2=0.38$, $p=0.082$). In the second randomised plot experiment, a significant difference ($p=0.05$) in both TP and reactive P (RP) forms in subsurface flow was recorded for different particle size fractions, with most TP transferred either in association with the 2-μm fraction or with the 0.001-μm or smaller fractions. Total P concentrations were higher from plots receiving P amendments compared with the zero-P plots, however, these differences were only significant for the >0.45-μm particle size fractions ($p=0.05$), and may provide evidence of surface applications of organic and inorganic fertilisers being transferred through the soil either as intact organic colloids or attached to mineral particles.

Conclusions Our results highlight the potential for drainage water to mobilise colloids and associated P during rainfall events, with wider implications for the transport of pathogens, fine sediment and persistent organic pollutants.

References

Heathwaite, A. L., P. M. Haygarth, R. Matthews, N. Preedy & P. Butler, P. (2005). Evaluating colloidal phosphorus delivery to surface waters from diffuse agricultural sources. *Journal of Environmental Quality* (in press).

Leaching losses of N, P and K from grazed legume based swards: some preliminary results

E.R. Dixon, A.C. Stone, D. Scholefield and D.J. Hatch
Institute of Grassland and Environmental Research, North Wyke Research Station, Okehampton, Devon, EX20 2SB, UK, Email: liz.dixon@bbsrc.ac.uk

Keywords: nutrients, leaching, grazing, legumes

Introduction There is increasing interest in sustainable agricultural systems because of environmental concerns. Animal production which utilises mixed grass and legume swards could be an effective measure in increasing the efficiency of nutrient utilisation, and investigation into different legume species is necessary. Leaching losses of N, P and K from 3 legume species under grazing by sheep were measured using *Teflon*-coated suction cups. The experiment took place on the UK site of the EU-funded, multi-site experiment – LEGGRAZE.

Materials and methods Three replicate plots (0.55 ha each) of 3 legumes species:- white clover (*Trifolium repens*), Caucasian clover (*Trifolium ambiguum*) and lotus (*Lotus corniculatus*), each in mixtures with ryegrass (*Lolium perenne L.*), were established in June 2002 at the Institute of Grassland and Environmental Research, North Wyke Research Station in Devon. The site is gently undulating on a free draining sandy loam soil (FAO dystric or eutric cambisol) over a deep water table. The plots were grazed by lambs (Suffolk x Mule) at a continuous variable stocking rate to maintain a sward height of approximately 7 cm, from mid August until mid-October. In October 2002, twenty *Teflon*-coated suction cups per plot were installed on a square (5 x 4) grid basis at a depth of 90 cm. These were sampled on 4 occasions during the drainage season (December to March). The samples were filtered (<0.45 microns) and frozen prior to analysis for NH_4^+-N, NO_3^--N, total N, soluble reactive P, and K. Drainage was calculated as: 'rainfall minus evapotranspiration' and each increment in drainage was allocated to the measured nutrient concentrations.

Results The total drainage for the season was 489 mm with 252 mm measured in December
N: White clover, lotus and Caucasian clover lost (mean values) 10.8, 8.2 and 6.4 kg inorganic N/ha and 13.4, 10.6 an 8.9 kg total N/ha, respectively, by leaching over the drainage season with greatest losses observed in December from all treatments. The strongest mean concentrations of NO_3^--N (mg/l) from each treatment were found in March: 2.9 (white clover), 2.4 (lotus) and 2.0 (Caucasian clover).
P: Caucasian clover, lotus and white clover lost (mean values) of 0.98, 0.48 and 0.47 kg soluble reactive P (SRP)/ha, respectively, over the drainage season, of which 0.66, 0.34 and 0.31 kg SRP/ha was lost in December. The strongest mean concentrations of SRP were found in March: 0.26 (Caucasian clover), 0.13 (lotus) and 0.12 (white clover) mg SRP/l.
K: Caucasian clover, lotus and white clover plots lost (mean values) of 22.0, 13.6 and 11.8 kg K/ha, respectively, over the drainage season of which 17.6, 7.3 and 7.1 kg K/ha was lost in December.
A visual assessment of percentage cover of the sown species conducted in September 2002 gave lotus; white clover and Caucasian clover mean covers of 22%, 20% and 7%, respectively.

Discussion and conclusions These preliminary findings show that, whilst the overall amounts of nutrients leached were relatively small, they are comparable with other studies. Drainage during the winter period was lower than average (*viz.* 89% of 30 year mean) which may have reduced the amount of nutrients lost. Mean NO_3^--N concentrations were weak relative to water quality legislative limits, but the highest mean SRP concentrations measured exceeded the OECD 35 μg/l total P threshold for increased risk of eutrophication in standing waters. The small differences in the amounts of N leached from the 3 legumes may be a reflection of the different legume contents of the swards, where interactions between source (N fixation) and sinks (unfertilised grass) and the extent to which the different legumes could have improved the soil structure (Scholefield, 2003; Rochon *et al.*, 2004 and Mytton *et al.*, 1993) would further complicate the between-species comparisons. However, these findings will be put into a wider environmental context when the results from all the European sites, over 3 years, are reported in a future paper.

References

Mytton, L. R., A. Cresswell & P. Colbourn (1993). Improvement in soil structure associated with white clover. *Grass and Forage Science,* 48, 84-90.
Rochon, J. J., C. J. Doyle, J. M. Greef, A. Hopkins, G. Molle, M. Sitzia, D. Scholefield & C. J. Smith (2004). Grazing legumes in Europe: a review of their status, management, benefits, research needs and future prospects. *Grass and Forage Science,* 59, 197-214.
Scholefield D. (2003). Some impacts of crop quality on environment and biodiversity. *Aspects of Applied Biology,* 70, 53-61.

Nitrogen dynamics following the break-up of grassland on three different sandy soils

M. Kayser, K. Seidel and J. Müller
Research Centre for Animal Production and Technology, University of Göttingen, Driverstrasse 22, D-49377 Vechta, Germany, Email: manfred.kayser@agr.uni-goettingen.de

Keywords: grassland break-up, Nmin, N mineralisation, hot water-soluble N

Introduction Nitrogen (N) is accumulated under grassland depending on factors such as soil type, management, and fertiliser input. Break-up of grassland stimulates mineralisation of organic N and may lead to increased soil mineral N and leaching losses (Lloyd, 1992). The objective of this study was to find out how site factors, e.g. soil, previous management and sward age, the following crop and the new level of N fertiliser affect the amount of inorganic N in autumn and the preceding mineralisation processes when grassland is ploughed in spring.

Material and methods The experiment had a three-factorial design with the main factors: site, crop following the break-up (barley + catch crop or maize), and level of N (i.e. 0, 120 or 160 kg/ha mineral N). All three sites are mainly sandy soils but of different origin and with differences in soil texture, organic matter, management and fertilisation history. Site 1, a plaggen soil, and site 2, a sandy podzol were mainly grazed and had little to moderate nutrient input. Swards were at least 15 years old, but the fields had been grassland for much longer. The third site is a deep-ploughed soil from shallow peat over sand and the 9-year-old sward had been used as cut grassland with high nutrient input via manures and mineral fertiliser. Among other characteristics, soil mineral N (Nmin), hot water-soluble N (Nhws) and N mineralised (Ninc), i.e. Nmin from *in situ* buried polyethylene bags minus Nmin at the start of incubation period, were analysed.

Results Results in the first year, after ploughing the grasslands in spring 2003, are shown in Table 1. Nmin in autumn and N leaching losses (not shown) were closely related. Differences between sites in Nmin were more pronounced than might be expected from Nhws or Ninc. Maize as a following crop stand seemed to stimulate the mineralisation more than barley, resulting in higher Nmin. The application of N fertiliser had no effect on Nhws or Ninc, but increased Nmin. N yields (not presented here) hardly differed between sites or level of N fertiliser. N mineralised (Ninc) from April to July corresponded well with Nmin in autumn (Figure 1).

Table 1 Means for hot water-soluble N (Nhws), N mineralised from April-July (Ninc), and Nmin in autumn

		Nhws (June) [mg/g soil]	Ninc (July) [kg/ha]	Nmin autumn [kg/ha]
Site	1	0.128 a	164 a*	117 a
	2	0.116 a	137 a	54 b
	3	0.103 b	164 a	134 a
Crop	Barley	0.109 b	128 b	53 b
	Maize	0.123 a	182 a	150 a
Fertilisation	N 0	0.113 a	154 a	74 b
	N 120/160	0.120 a	156 a	129 a

* values with different letters are significantly different at the P<0.05 level

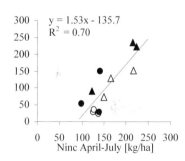

Nmin autumn [kg/ha]

$y = 1.53x - 135.7$
$R^2 = 0.70$

Ninc April-July [kg/ha]

Figure 1 Relationship between N mineralised April-July (0-30 cm; Ninc) and Nmin autumn (0-90 cm). Barley = ●; maize = ▲; filled = N 120 (barley) or N 160 kg N/ha (maize); empty = no N

Conclusions The results demonstrate the strong effect of the stimulated N mineralisation after ploughing. Hot water-soluble N (Nhws) and N from incubation (Ninc) seemed to be similarly influenced by crop stand and fertilisation, though they only partly accounted for differences in Nmin between sites or level of N fertilisation. It is recommended that barley be grown instead of maize as a following crop and to minimise N fertilisation to reduce the risk of N leaching.

References
Lloyd, A. (1992). Nitrate leaching under arable land ploughed out from grass. *The Fertilizer Society*, 330, 1-32.

Mechanical aeration and liquid dairy manure: application impacts on grassland runoff water quality and yield

T.J. Basden, S.B. Shah and J.L. Miller
West Virginia University Extension Service, P.O. Box 6108 Morgantown WV, 26506 USA, Email: tom.basden@mail.wvu.edu

Keywords: aerator, simulated rainfall, nitrogen, phosphorus, total suspended solids

Introduction Wet weather on heavy soils reduces oxygen availability in the root zone and reduces forage yields. Mechanical aeration can improve forage yield in these soil conditions. Research has shown that under certain conditions, mechanical aeration can increase yield by improving drainage and aeration (Davies *et al.*, 1989); aeration can also increase depression, storage and infiltration thus reducing surface runoff and improving nutrient distribution in the root zone. Aeration on sloping, fertilised grassland can provide environmental (Douglas *et al.*, 1995) and agronomic benefits. The objectives of this study were to evaluate the runoff water quality and agronomic impacts of mechanical aeration and liquid dairy manure (LDM) applied to hillside grasslands.

Materials and methods A mechanical Aerator (Model: AerWay®) was evaluated on pasture consisting of cocksfoot with 10-20% lucerne, which had not received LDM for 6 years. The soil is a well-drained silt loam. The experiment was RCB with four treatments and three replicates. Each treatment block (2.5mx 2.5m) received 67 mm of simulated rainfall (SR) to generate runoff. Each block was surrounded on three sides with metal borders and on the fourth (down-slope) side with a runoff collector. Treatments included: 1) control, no aeration and no liquid dairy manure (CTL), 2) aeration only (AER), 3) manure only (MAN), 4) aeration and manure (AER+MAN). Runoff water quality analysis included: nitrate, ammoniacal-N, Total N (TKN), Dissolved Reactive Phosphorus (DRP), total P, Total Suspended Solids (TSS), runoff depths and rainfall leaving plots.

Results Water quality impacts included 1) Runoff Depth; Aeration during spring did not improve infiltration of water. 2) Nutrients; Nutrient concentrations in the simulated runoff events were higher with LDM and were unaffected by aeration. Aeration reduced losses of three or more nutrient species (N and P) in two of six SR events only in manured plots. Concentrations and loadings indicated that aeration of manured plots was more effective in reducing DRP losses than other species (Table 1). Total mean loadings of individual nutrient species in SR were reduced by $\geq26\%$ by AER+MAN vs MAN. 3) Suspended Solids; TSS concentrations were significantly higher with aeration but not with LDM application. Loadings of TSS from AER were >30% higher than the other three treatments. Forage yield impacts 1) In two of three harvests, MAN increased forage yield significantly vs CTL and AER while AER+MAN reduced yield vs MAN in one harvest. 2) Compared with MAN, total forage yields with CTL, AER, and AER+MAN were 78%, 67%, and 81%, respectively.

Table 1 Pollutant loadings in the first simulated runoff event

Treatment	Mean (SD)[a] nutrient loading (g/ha)					Mean (SD)[a] TSS loading (kg/ha)
	NO$_3$-N	Ammoniacal-N	TKN	DRP	Total P	
CTL, control	27 (17)	4c[b] (4)	269 (101)	23(9)	32b (10)	2.9 (1.1)
AER, aeration	19 (3)	1c (2)	254 (92)	45 (55)	27b (4)	1.8 (0.6)
MAN, manure	40 (18)	112a (18)	531 (24)	210 (20)	186a (21)	2.4 (2.0)
AER+MAN Aeration+Manure	40 (17)	72b (14)	644 (347)	151 (160)	181a (102)	4.8 (2.4)
LSD[d]	(NS)	17	(NS)	(NS)	79	(NS)

[a]Mean and standard deviation based on three replicates;[b]Treatment means, followed by the same letter are not significantly different at α=0.05;,[c]ANOVA; [d]Fisher's least significant difference (p=.05)

Conclusions Aeration partially improved runoff water quality from manured grassland but adversely affected crop yield and nutrient uptake. Further studies should be performed on pastures with livestock traffic. Also, there is need for aerators that minimise surface soil disturbance to reduce TSS losses.

References
Davies, A., W. A. Adams & D. Wilman (1989). Soil compaction in permanent pasture and its amelioration by slitting. *Journal of Agricultural Science*, 113, 189-197.
Douglas, J. T., C. E. Crawford & D. J. Campbell (1995). Traffic systems and soil aerator effects on grasslands for silage production. *Journal of Agricultural Engineering Research*, 60, 261-270.

Management options to reduce N-losses from ploughed grass-clover

J. de Wit, G.J. van der Burgt and N. van Eekeren
Louis Bolk Institute, Hoofdstraat 24, 3972 LA Driebergen, The Netherlands, Email: j.dewit@louisbolk.nl

Keywords: nitrate leaching, N-losses, grass-clover, silage maize

Introduction Nitrate (NO_3^-) leaching from grassland can be kept at acceptable levels, but is often high after ploughing for grassland renewal or for silage maize/grain production. In on-farm research with several organic farmers, management options are being explored to save scarce organic manure and to reduce N-losses.

Materials and methods A trial was conducted on a loamy löss (2.6% OM in 0-30 cm) with four treatments: A) 'standard' farmer practice, i.e. application of $20m^3$ of slurry in early spring (21 April), mowing grass (26 May), soil ripping, application of $18m^3$ slurry, ploughing and sowing of maize (1 June); B) and C) were similar to A, but with no slurry application before ploughing, or no slurry at all, respectively. D) as C), but with early soil ripping (27 April) to enhance N-availability. Field history included three years of arable crops followed by one-year grass-clover (> 50% clover). Mineral N-availability was measured 9 times at regular intervals, maize production was assessed by harvesting three rows of 3 m/plot. Mineral N-availability in the layer of 0-30 cm was modelled by the soil-N flow model NDICEA (Koopmans & Bokhorst, 2002).

Results The recorded mineral N-availability was lower than standard Dutch advice (185 kg N/ha), while recorded residual N was higher than the maximum Dutch advice to attain NO_3 leaching <50 mg/l, i.e. 90 kg N (Table 1). The recorded production follows mineral N-availability reasonably closely except for the early soil ripping treatment. Figure 1 shows the NDICEA-results, matching the recorded mineral N-levels closely (Table 2) except for the late growing period, possibly due to underestimation of the capillary capacity resulting in a predicted decomposition rate being lower than reality (2003 was an extremely dry year). Figure 1 also shows NDICEA-results of "treatment E", being similar to D but with silage maize following three-year-old grass-clover. Mineral N-availability is predicted to be much higher due to the decomposition of a much higher amount of easily available soil OM (roots and living soil organisms).

Table 1 Maize production and mineral availability for different management options

Treatment	Maize production (ton DM/ha)	Mineral N (kg/ha) at: 21-6 (0-60 cm)	25-11 (0-90 cm)
A	16.8	136	135
B	15.8	133	100
C	15.3	113	58
D	16.1	175	103

Table 2 Recorded mineral N-availabity (0-30 cm, kg/ha) for different treatments at different dates

	23-4	7-5	6-6	21-6	3-9	25-11
A	25	23	83	107	65	58
B	20	18	89	110	26	46
C	13	15	71	88	24	37
D	10	66	121	143	77	49

Conclusion Silage maize production following grass-clover was increased by manure application but N-losses, both residual- N and denitrification losses (not shown), increased much more. This is particularly true for grass-clover leys older than 2 years. Early soil ripping enhances mineral N-availability and silage maize production without applying scarce organic manure, but also reduces grass production and increases potential N-losses. Therefore, future experiments will be concentrated on maize production without destroying the grass-clover to facilitate a more direct transfer of fixated N from clover to maize and possibly lower inherent N-losses.

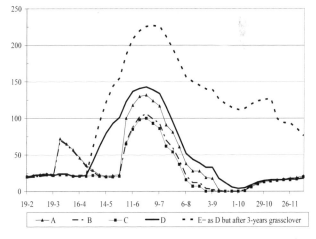

Figure 1 Simulated mineral N availability (kg/ha, 0,30cm)

References

Koopmans, C. J. & J. Bokhorst, 2002. Nitrogen mineralisation in organic farming systems: a test of the NDICEA model. *Agronomie* 22, 855-862

Rangeland ecological management counter-measures study of Xinjiang

H.X. Cui, J. Li, S. Asiya, J.L. Zhang and Jialin
Grassland Research Institute of Xinjiang Animal Science Academy, Urumqi, 830000, China, Email: jie1442@sina.com

Keywords: grassland, treatment and counter-measures, environmental capacity

Introduction Xinjiang is located in the hinterland of the Asian-Euro continental area and far from the ocean. It is surrounded by mountains and forms a physiognomic landscape of two basins located in three mountains. This results in an extremely droughty continental climate and a harsh environment in the Temperate Zone. However, the geographical environment is diverse in the Xinjiang Autonomous region, the three high mountains and complicated geographical structure results in changes in ecological conditions with altitude and a diverse ecological environment. Because of the drought characteristics the desert area has developed different types of vegetation, and the complexity of the system is increased and the rangeland plays a very important role in social economic development in the Xinjiang Area.

The main issues in the management of the rangeland are: 1) because of agricultural development, a large area of pasture was utilised and soil water supplies have been jeopardised: the yield of pastures have been reduced but the number of animals have increased. 2) the harsh environment and the poor management system are the major problems of rangeland management in Xinjiang. The herdsmen in Xinjiang rely on natural conditions to supply sufficient feed for their animals, so the ability to sustain production when conditions are extreme are lowered and reduce the ability to supply feed during the long winter.

Methods The large area of rangeland, abundant plant varieties and diversity of grassland which, with better management, could not only provide the foundation for improved animal husbandry but also be an important buffer for the ecological security of Xinjiang. However, because of long term over-grazing and rangeland misuse, the rangeland has degenerated seriously. The study aims to optimise the important resources of the rich soil and available water to establish renovated pasture, to enable herdsmen to employ a half year yard feeding system. We wish to improve forage resources in order to support the implementation of the 'Return Grazing Land to Pasture Project'. The aims also include the permanent closure of the desert-pasture and poor productivity desert-pasture which is not suitable for grazing; re-investigate and confirm grazing rates for degenerated grassland to reduce grazing pressure and protect grassland. Advice will be based on rates of pasture self-renovation to rebuild degenerated grassland. We will determine the pastures' environmental capacity, and convey information on ecological transfers to herdsmen to achieve a balance between pasture productivity and the number of animals.

Aims and conclusions Our studies will provide the means of determining the requirements for returning the grazed land into pasture and recovering the degenerated grasslands. By developing improved pasture, we aim to change the current pattern of a whole year's grazing system to one which will involve a yard feeding system for half of the year. The pasture area's environmental capacity will be determined and the importance of this demonstrated to the herdsmen so that they can establish a system which allows a balance between the pastures' productivity and stability and the number of livestock. This will have to be established in accordance with regulations.

Green Dairy, a project for environmental friendly and sustainable dairy systems in the Atlantic area

H. Chambaut[1], A. Pflimlin[2] and C. Raison[3]

[1]Institut de l'Elevage, Angers, France, Email: helene.chambaut@inst-elevage.asso.fr, [2]Institut de l'Elevage, Paris, France, [3]Raison C, Institut de l'Elevage, Le Rheu, France

Keywords: pilot farms, nutrient supply, losses to the environment

Introduction Green Dairy is an European Interreg III B project which runs over 3 years. It brings together 10 research and development partners from 5 countries and 11 regions of the Atlantic Area which range from Scotland to Portugal. It aims to provide a better understanding of the impact of intensive dairy systems on the quality of the environment in order to develop ways of improving practices in the different regions. This is expected to encourage more rapid responses to the problems of deteriorating water and air quality and a more appropriate response within or between regional contexts that could be used as proposals for implementation of regulations or advice.

Materials and methods Well motivated farmers have been selected to build a network of pilot dairy farms which aims to test different ways of optimising practices to reduce nutrient losses from the farming systems. The systems studied cover a wide range of levels of intensification whether expressed on a per cow, per hectare or even per worker basis. The changes expected will also have to integrate the local geographical and economic context of each region in which they are situated. Thus, dairy farm systems can be based on the maximum use of grazing on the one hand or on maize silage only with high quantity of slurry to spread on the other. This project aims to develop discussions on the diverse ways of optimising each farming system according to the local environment and regulations. At the same time, a network of research stations has been established in which mineral nutrient flows are monitored on complete dairy systems in order to make possible the identification of the critical points for each specific problem and also to measure the effectiveness of various options to, for example, protect water resources. Mathematical modelling will be used to describe nutrient flows and losses at the entire farm system. These results will then be used to provide options for the farmers to increase their nutrient use efficiency and to have a better understanding of the controls over water quality that are observed in the Atlantic area by undertaking modelling and mapping studies. A final seminar is planned in September 2006 at which the final results of this study (which involve 5 countries and 10 partners: 1) United Kingdom, ARINI, N. Ireland; IGER, Devon and Wales; SAC, Scotland: 2) Ireland, TEAGASC, Moorepark, SW Ireland: 3) France, Institut de l'Élevage, CRAB Brittany, CA44 Pays de la Loire and Aquitaine: 4) Spain, CIAM, Galicia; Neiker, Basque Country: and 5) Portugal, UTAD, North Portugal) will be presented. Further details of the project are given at the Green Dairy (2005) website.

Results The preliminary results are on the characterisation and on the initial environmental assessment and diagnosis of the farming systems. Mineral balances at the farm gate scale calculated for the commercial pilot farms shows that the excess of nitrogen (N) varied from 100 to 500 kg N/ha of the agricultural area, from 13 to 29 kg N/1000 l of milk sold and the efficiency of N utilisation rate from 29 to 38%. Nitrate concentration in drainage water varied from 10 to 50 mg NO_3 /l: at present a direct link to the contributions of the various dairy farming activities in the total land use of the regions and of natural conditions (annual rainfall, winter drainage, denitrification etc.) has not been determined. At the experimental sites, mineral flows within the farming systems have been determined during 2004 and relationships with N soil contents measured before the oncoming drainage period. Estimation of nitrate losses by measurement of nitrate concentration in drainage water or by modelling will provide a means of determining the proportion of the excess of the farm gate balance that is expected to be lost for each system under the various conditions of the Atlantic region.

Conclusions The present study will provide practical information on opportunities to increase the efficiency of nutrient use and reduce pollution throughout the wide ranging conditions of the Atlantic region. As well as determing new scientific information, the nature of the programme means that there are direct interactions between farmers and scientists to enable knowledge to be effectively and directly transferred from research to farm and *vice versa*.

References
Green Dairy (2005). www.inst-elevage.asso.fr/greendairy/

SAFE - a tool for assessing the sustainability of agricultural systems: an illustration

X. Sauvenier[1], C. Bielders[2], M. Hermy[3], E. Mathijs[4], B. Muys[3], J. Valckx[3], N. Van Cauwenbergh[2], M. Vanclooster[2], E. Wauters[4] and A. Peeters[1]

[1]Laboratoire d'Ecologie des Prairies, UCL, Croix du Sud, 5 bte 1, 1348 Louvain-la-Neuve, Belgium, Email: sauvenier@ecop.ucl.ac.be, [2]Unité de Génie Rural, UCL, Croix du Sud, 2 bte 2, 1348 Louvain-la-Neuve, Belgium, [3]Laboratorium voor Bos, Natuur en Landschap, Vital Decosterstraat, 102, 3000 Leuven, Belgium, [4]Afdeling Landbouw- en Milieueconomie, KUL, Willem de Croylaan, 42, 3001 Leuven, Belgium

Keywords: sustainability, integration, framework, analysis,

Introduction SAFE (Framework for Assessing Sustainability levels) is a tool for evaluating the sustainability of agricultural systems and uses a hierarchical framework populated with indicators objectively selected by multi-criteria evaluation. Indicators are measured at field, farm and landscape scales and progressively integrated into a global sustainability index (SI). SAFE is illustrated below with results on a field scale from a farm site.

Results The outcomes of this study are shown in Tables 1 and 2 and Figure 1.

Table 1 Principles and criteria of the SAFE hierarchical framework for soil resources

ENVIRONMENTAL PILLAR: soil resource	
Principles	Criteria
	Soil loss is minimised
Soil regulation function shall be maintained or enhanced	Soil chemical quality is maintained or increased
	Soil physical quality is maintained or increased

Table 2 Potential indicators, results of selection and 'fuzzification' of selected indicators

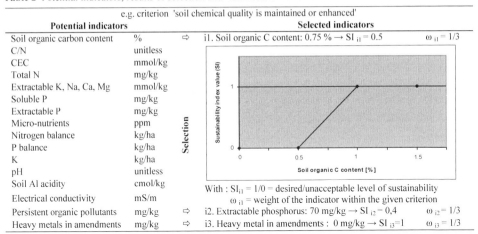

e.g. criterion 'soil chemical quality is maintained or enhanced'

Potential indicators			Selected indicators	
Soil organic carbon content	%	⇒	i1. Soil organic C content: 0.75 % → SI $_{i1}$ = 0.5	ω_{i1} = 1/3
C/N	unitless			
CEC	mmol/kg			
Total N	mg/kg			
Extractable K, Na, Ca, Mg	mmol/kg			
Soluble P	mg/kg			
Extractable P	mg/kg			
Micro-nutrients	ppm			
Nitrogen balance	kg/ha			
P balance	kg/ha			
K	kg/ha			
pH	unitless			
Soil Al acidity	cmol/kg			
Electrical conductivity	mS/m		With : SI $_{i1}$ = 1/0 = desired/unacceptable level of sustainability ω_{i1} = weight of the indicator within the given criterion	
Persistent organic pollutants	mg/kg	⇒	i2. Extractable phosphorus: 70 mg/kg → SI $_{i2}$ = 0,4	ω_{i2} = 1/3
Heavy metals in amendments	mg/kg	⇒	i3. Heavy metal in amendments : 0 mg/kg → SI $_{i3}$=1	ω_{i3} = 1/3

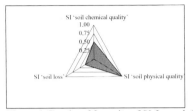

⇒ Integration of indicators at the 'criterion' level:

$$SI_{soil\ chemical\ quality} = SI_{i1}*\omega_{i1} + SI_{i2}*\omega_{i2} + SI_{i3}*\omega_{i3} = 0,63$$

Further integration (from the 'principle' to the 'global' level) requires weighing defined by the end's user. Results at the principle level are displayed with spider-web graphs in Figure 1

Figure 1 Results of farm site of SI for each 'criterion' related to the 'principle' *soil regulation function*.

Conclusions The 'sustainability index' function related to an indicator is case specific: the shape is based on expertise and support points either as reference values, expertise or minimum/maximum/average values taken by the indicator in similar contexts. Indicator weightings (ω) within a given criterion are extrapolated from their respective 'relevance to sustainability' scores given by experts during the multi-criteria evaluation.

SAFE - a tool for assessing the sustainability of agricultural systems: methodology

X. Sauvenier[1], C. Bielders[2], M. Hermy[3], E. Mathijs[4], B. Muys[3], J. Valckx[3], N. Van Cauwenbergh[2], M. Vanclooster[2], E. Wauters[4] and A. Peeters[1]

[1]Laboratoire d'Ecologie des Prairies, UCL, Croix du Sud, 5 bte 1, 1348 Louvain-la-Neuve, Belgium, Email: sauvenier@ecop.ucl.ac.be, [2]Unité de Génie Rural, UCL, Croix du Sud, 2 bte 2, 1348 Louvain-la-Neuve, Belgium, [3]Laboratorium voor Bos, Natuur en Landschap, Vital Decosterstraat, 102, 3000 Leuven, Belgium, [4]Afdeling Landbouw- en Milieueconomie, KUL, Willem de Croylaan, 42, 3001 Leuven, Belgium

Keywords: sustainability, hierarchical framework, multi-criteria analysis, integration

Introduction Issues of sustainability are challenges for scientists, politicians and farmers, generating many sets of indicators at national and international levels. However, most initiatives focus only on environmental aspects; indicators are often selected arbitrarily and do not fit in a consistent, comprehensive and universal framework. There is a need to integrate data to facilitate comparison and diagnosis. SAFE (Framework for Assessing Sustainability levels), a tool for evaluating sustainability of agricultural systems was developed in this context.

Methods and results SAFE defines three hierarchical levels 'principles, criteria and indicators' (Table 1) for each pillar of sustainability (environmental, social and economic). These reflect the multi-functions that agricultural systems should provide to be sustainable. A list of potential indicators based on a literature review was submitted to experts for multi-criteria evaluation (7 expertise selection criteria or 'ESC') (Table 2). Selection for this was based on averages of expert scores (Sen, 1986) for each ESC: the 'agreed ESC score'. Indicators were selected if (1) the 'agreed ESC score' for ESC 'relevance to sustainability' was greater than a minimum threshold and (2) the average of the seven 'agreed ESC scores' was in the given range for the best indicators related to the same sustainability criterion. This determined a core set of relevant and practical indicators that were measured and/or calculated at field, farm and landscape scales. Fuzzy logic (Cornelissen, 2003) transformed each value into a 'sustainability index' (SI) whose scale $[0 \rightarrow 1]$ is continuous and unit less. This permits integration of indicators related to the same sustainability criterion; the weighing being extrapolated from the 'relevance to sustainability' 'agreed ESC score' of each one. Further integration required weighting by the end user. The results were represented as a 'web'. SAFE was tested on four farms

Table 1 Hierarchical framework ('P, C &I')   **Table 2** Expertise selection criteria (ESC)

GOAL Sustainable agriculture integrating environmental, economic and social aspects	**1. Discriminating power in time and space** → Ability to discriminate between changes due to external factors and management, in space and time
	2. Analytical soundness → Is the indicator scientifically valid?
PRINCIPLE General conditions for achieving sustainability, relating to the multifunctionality of agro-ecosystems *Example: Soil regulation function of the agro-ecosystem shall be maintained or enhanced*	**3. Measurability / Cost and time involved** → Is the use of the indicator justified in terms of cost and time consumption?
	4. Transparency *How understandable is the indicator?*
CRITERION Resulting state of the agro-ecosystem when a principle is respected. *Example: Soil chemical quality is maintained*	**5. Policy relevance** → Does the indicator help in monitoring effects of polcies and identifying areas where action is needed?
	6. Transferability → Does the indicator relate to general practices of major farm types?
INDICATOR Quantitative or qualitative variable which can be assessed in relation to a criterion. *Example: CEC*	**7. Relevance to sustainability criteria** → Is the indicator a relevant measuring tool for the sustainability criterion/criteria it is related to?

Conclusions SAFE uses a structured framework to evaluate the sustainability of agricultural systems at different scales. This framework is populated with indicators objectively selected (multi-criteria evaluation by experts). SAFE integrates progressively the information for selected indicators.

References

Cornelissen, A. M. G., J. van den Berg, W. J. Koops, M. Grossman & H. M. J. Udo (2001). Assessment of the contribution of sustainability indicators to sustainable development: a novel approach using fuzzy set theory. *Agriculture, Ecosystems and Environment*, 86, 173-185.

Sen, A. (1986). Social Choice Theory. *In*: K. J. Arrow & M. D. Intriligator (eds) *Handbook of Mathematical Economics, vol III*. Elsevier Science Publishers B.V., North Holland. pp. 1073-1181.

Keyword index

Author index